M000314101

Praise The Simulated Multiverse Hypothesis

"How am I unreal? In *The Simulated Multiverse*, Riz Virk counts the ways. So many ways we might all live in one (or more) simulated worlds. Palpable 'reality' may be as delusional as our old notion that the heavens revolved around Earth. And just like Galileo did then, Virk may help open our eyes to a greater (if humbling) cosmology."

—**David Brin**, author of *EXISTENCE*, *The Postman* and *EARTH*.

"A few cutting-edge thinkers have posited that this world in which we live is a simulation. In *The Simulated Multiverse*, his second book on the subject of simulation, Rizwan Virk walks us through the concept of a computer-generated reality that Philip K. Dick posited in the speech that he delivered in Metz, France, in 1977. This volume is well-researched and eminently readable. Perhaps *The Matrix* is more than a science fiction film; perhaps it reveals some truth about our world and our lives in this world. Read this book, if you dare."

— **Tessa B. Dick**, wife of Philip K. Dick and author of *Conversations with Philip K. Dick,*

"Refining his impressions first introduced in *The Simulation Hypothesis*, with the progress of AI since his research at MIT, in *The Simulated Multiverse*, Virk has expanded his thesis in the wider scope of the multiverse, building on the framework of pioneers such as Nick Bostrom and Kurzweil. The result is a new challenge not only to theories of simulation, but to what constitutes reality itself, and human illusions of our rightful place within it."

— **Jacques Vallée**, venture capitalist, author of *Forbidden Science*

"Virk ... makes a cogent, clear-eyed guide to the head-spinning science of parallel universes, quantum indeterminacy, and the possibility— terrifying or relieving—that our perceived reality is in fact part of a great simulation."

—***Publishers Weekly***

"I'm grateful to be living in a branch of the multiverse where if I'm asked to go into more detail about the relationship between *Total Recall* and quantum physics, I can simply point someone in the direction of this comprehensive and entertaining book. To quote a certain 1999 blockbuster: Woah."

—**Rodney Ascher**, director of *A Glitch in the Matrix*

"In *The Simulated Multiverse*, Riz Virk takes simulation theory and *The Matrix* to a new level. Using computational tools – complexity, artificial intelligence, video games and quantum computing –Virk explains an interpretation of our world that sounds like science fiction: that we are living multiple parallel timelines. If you want to get glimpses of possible presents, pasts, and futures, read this book!"

‑*Brad Feld*, venture capitalist and author of
The Startup Community Way

"MIT scientist's '*Simulation Hypothesis*' makes compelling case for The Matrix."

—*The Next Web*

"My own experience has taught me that we live, teach, learn and love in a virtual world. In *The Simulation Hypothesis*, Riz Virk combines the mind of a scientist with the heart of a mystic, using video games to explain the virtual reality that we live in."

—**Dannion Brinkley**, bestselling author of *Saved by the Light*
and *At Peace in the Light*

"In *The Simulation Hypothesis,* Riz Virk provides a deft and knowledgeable blend of video game history, hard science speculation, and science fiction reference ... I found it fascinating and entertaining."

— **Noah Falstein**, former chair of the IGDA, former Chief Game
Designer at Google

"Rizwan Virk's book *The Simulation Hypothesis* is one of the few works that could convince me that I probably live in a simulated universe. If this sounds mind blowing, it is!"

—**Diana Walsh Pasulka**, author of *American Cosmic: UFOs,
Religion, Technology*

Praise for *Zen Entrepreneurship* and *Treasure Hunt*, also by Rizwan Virk:

"*Treasure Hunt* provides some well-worth-your-consideration guidance. Virk provides new maps of understanding, based on the latest thinking of quantum physics and the multiverse, that can guide you through today's jungle of opportunities and dense misadventures."

—— **Fred Alan Wolf PhD**, author of *Parallel Universes* and *Dr. Quantum Presents: Do-it-Yourself Time Travel*

"So exciting that someone from the tech world is speaking up about synchronicities, signs, and spiritual guidance. Thanks, Riz, for giving us case studies and a compelling guide for finding the map we all contain inside."

—— **Pam Grout**, #1 New York Times bestselling author of *E-squared*

"The world around us is speaking to us every day in a language of signs and symbols, if only we pay attention. Virk invites us to look at the patterns of everyday life as a treasure map, offering clues we can follow to manifest our dreams."

—— **Robert Moss**, author of *Conscious Dreaming* and *Sidewalk Oracles*

"*Tales of Power* meets *The Peaceful Warrior*... in Silicon Valley! *Zen Entrepreneurship* is entertaining, humble, insightful and valuable—not just to entrepreneurs, but to anyone looking to manifest their dreams and make a difference in the world."

—— **Foster Gamble**, Creator and Host, *Thrive: What on Earth Will It Take*

"In *Zen Entrepreneurship*, Riz Virk brings the wisdom of ancient Eastern traditions into a purely Western setting. The result is an often hilarious but always insightful book that will change how you view career success and help you discover and walk your own unique path."

—**Marc Allen**, author of Visionary Business, CEO and co-founder of New World Library

ALSO BY RIZWAN VIRK

The Simulation Hypothesis:
An MIT Computer Scientist Shows Why AI, Quantum Physics and Eastern Mystics Agree We Are in a Video Game

Startup Myths & Models:
What You Won't Learn in Business School

Treasure Hunt:
Follow Your Inner Clues to Find True Success

* Wisdom of a Yogi:
Lessons for Modern Seekers from Yogananda's Autobiography

THE ZEN ENTREPRENEUR SAGA

Zen Entrepreneurship:
Walking the Path of the Career Warrior

* The Zen Entrepreneur and the Dream:
Further Adventures in Business and Consciousness

visit
www.zenentrepreneur.com

* = COMING SOON

The Simulated Multiverse

An MIT Computer Scientist Explores
Parallel Universes, the Simulation Hypothesis,
Quantum Computing and
the Mandela Effect

Rizwan Virk

v.8.9

BAYVIEW

BOOKS

www.bayviewlabs.com/bayviewbooks/

Publisher's Cataloging-in-Publication data

Names:	Virk, Rizwan, author.
Title:	The simulated multiverse : an MIT computer scientist explores parallel universes, the simulation hypothesis, quantum computing and the Mandela Effect / by Rizwan Virk
Description:	ISBN: 978-1-954872-00-4 (paperback) | 978-1-954872-01-1 (ebook) | 978-1-954872-02-8 (ISBN/kindle), B08XFR749T (ASIN/kindle) | LCCN: 2021913840
Identifiers:	ISBN: 978-1-954872-00-4 (paperback) | 978-1-954872-01-1 (ebook) | 978-1-954872-02-8 (ISBN/kindle), B08XFR749T (ASIN/kindle) | LCCN: 2021913840
Subjects:	LCSH: Multiverse. | Space and time. | Computer simulation. | Virtual reality. | Quantum theory. | Video games. | Quantum computing. | Computational complexity. | Cellular automata. | Dick, Philip K.--Influence. | Time travel. | Quantum cosmology. | Artificial intelligence. | Philosophy of mind. | Near-death experiences. | Computers and civilization.
Classification:	LCC: QB991.Q36 V573 2021 | DDC: 523.1--dc23

For our nephews and nieces,
who easily grasped the multiverse

Danny
Gianna
Daniyal
Rayyan
Emaad
Taimur
and
those to come

Table of Contents

Part I

Sounds Like Science Fiction

"People assume that time is a strict progression of cause to effect, but...it's more like a big ball of wibbly-wobbly, timey-wimey stuff."

—The Doctor, Doctor Who[1]

"Can't repeat the past?" he cried incredulously.

"Why of course you can!"

—F. Scott Fitzgerald, The Great Gatsby

Chapter 1

Down the Rabbit Hole—From Google into the Mind of Philip K. Dick

We are living in a computer-programmed reality, and the only clue we have to it is when some variable is changed, and some alteration in reality occurs. We would have the overwhelming impression that we were re-living the present—déjà vu—perhaps in precisely the same way: hearing the same words, saying the same words.

-Philip K. Dick, *Metz Sci Fi Convention 1977*[2]

This book is about a complex idea that may sound like science fiction: that we live inside a simulated multiverse. In case you aren't familiar with this idea, it is built on top of two conclusions that, though they might seem fringe, are increasingly supported by many scientists, philosophers, and religious scholars.

The first is that we live inside a digital, simulated world, a high-resolution video game that is similar to the world depicted in the blockbuster movie, *The Matrix*. This concept is broadly referred to today as the simulation hypothesis, and it was the subject of my previous book of that name. It implies that the three-dimensional world around us (what we call space) is not what we think it is.

The second is that far from living in a single universe, we live in a complex, interconnected network of multiple timelines. This concept is broadly referred to today as the multiverse. Not only does the multiverse warp our understanding of the world around us, it also warps our understanding of the past and the future. In short, neither space nor time is what we think it is.

We will explore many other concepts in this book that support these conclusions—including quantum indeterminacy, quantum computing, video game design, and the Mandela effect. But before we get into the details, I wanted to tell you a bit about my journey from a video game entrepreneur and creator of a virtual reality program at MIT, down the rabbit hole of simulation theory.

From Ping-Pong to *The Matrix*

You could say that I have been obsessed with science fiction and computers my whole life, and not surprisingly, it was the intersection of these two fields that got me started thinking about the simulation hypothesis. This in turn led me to thinking about the simulated multiverse.

A few years before publishing *The Simulation Hypothesis*, in 2016, I had just sold my last video game company and was wondering what to do next with my life. I visited a startup that was building virtual reality (VR) games. VR had captured Silicon Valley's mantle of the *next big thing*. Facebook had bought Oculus for $2 billion recently, and other technology giants like HTC and Sony were throwing their hats into the virtual reality ring with their own VR headsets.

I visited this startup's office in Marin County, across the Bay from the city of San Francisco, and tried out their new sports VR game. It was a room-scale setup, which means the room was pretty much empty, except for a computer in the corner linked to some wires hanging down from the ceiling. Most of the room was a taped-off square area that served as the arena where the VR player could move around freely. I put the headset on and looked out across the virtual landscape; I saw a virtual ping-pong table and a virtual opponent.

A paddle magically appeared in front of my hand (which in reality was holding the controller), and as I moved my hand, the paddle moved. Suddenly, a ball appeared, and I started playing against my virtual opponent. Over the next few minutes, I became completely engrossed in the virtual table tennis game. The responsiveness of the system and its underlying physics engine were perfect; it felt like my paddle was hitting a ball, and the ball was following a natural trajectory to bounce off the table toward my opponent. I became so lost in the illusion that by the end of the game, I instinctively put the "paddle" down onto the "table" and attempted to lean on the table, just like I might do after a real table tennis game.

Of course, there was no paddle and no table. The controller in my hand fell on the floor, and I almost fell over as I tried to lean on the nonexistent table. That's when I realized that VR had started to achieve the kind of immersion that could fool the human mind.

Studies have shown that the brain responds to perceived stimuli in a virtual environment in the same way as it does to real stimuli in a physical environment. For example, if you are standing on the roof of a tall building in VR and you are afraid of heights, you start to have similar physiological responses. Companies have used this knowledge to use VR as an effective therapy to overcome fear of things like heights or spiders, all of which can be simulated safely inside virtual reality.

The virtual ping-pong experience, which I have spoken about many times, was one of several VR experiences that led me to wonder about immersive simulations. Later that same year, I donned another VR headset and found myself in a virtual cavern, standing on a virtual ledge next to a very steep drop into what looked like the bottomless chasm in the mines of Moria in *The*

Lord of the Rings. Despite having the intellectual knowledge that I wasn't really in the cavern and wasn't in any danger, my body was afraid to move my foot two steps to the side for fear of falling into the dark depths below.

These experiences led me to wonder what elements would have to be in place for us to build a world that was, for all practical purposes, indistinguishable from physical reality and how long it would take our technology to get there.

In my previous book, *The Simulation Hypothesis*, I laid out a roadmap of stages of technology, starting with simple video games and ending up with fully immersive virtual-world simulations that were as convincing as those in *The Matrix*. This would take us to a theoretical point in the future that I like to call the Simulation Point.

I concluded that we weren't that far from the Simulation Point. To my surprise, there was a well-known argument by Oxford philosopher Nick Bostrom (made in his 2003 paper, "Are You Living in a Computer Simulation?") that if any technological civilization ever reached the Simulation Point, we were almost certainly living in a simulation ourselves. Although this sounds like an odd argument at first glance, it has gotten more and more support over time, and I will revisit it in Chapter 3. It turns out that Bostrom wasn't the first philosopher to tell us that the world around us may not be real, and we'll dive into some of these in that chapter also.

Surely, the physicists would be able to give us more confidence that the world around us is a physical construct, I thought. Yet, even more surprising to me was that when I explored some of the biggest mysteries in the world of physics, I found that they could be much more easily resolved if we were living in a simulated reality and not in a purely physical reality. In fact, I found that many prominent physicists had reached the conclusion that the physical world consisted not of physical matter but of information, a conclusion that formed the underpinnings of *The Simulation Hypothesis.*

Moreover, moving beyond computer science and physics and philosophy into the realm of religion, I realized that this idea had been a key idea not of any one religion, but of all the world's religions—including Eastern religions like Buddhism and Hinduism and the Western religions of Judaism, Christianity, and Islam.

Having written *The Simulation Hypothesis* and explored this idea in depth from all of these angles, I was satisfied that I had been down the rabbit hole and was ready to emerge and resume my career in Silicon Valley and in academia.

It was then that I had several unexpected conversations whose implications caused me to reconsider the width and depth of the rabbit hole. The implications, which I couldn't quite shake, included the thought that if one timeline could be simulated, there was no reason that multiple timelines couldn't be simulated using the same "computer system." Each simulated timeline would basically be a different run of the simulation, with some variables changed. This led me down a winding but scenic road of curiosity from Google into the mind of famed science fiction writer Philip K. Dick and into the quantum world, eventually to settle on the core idea of this book: that we live in a *simulated multiverse.*

Near the Googleplex

Not long after I had published *The Simulation Hypothesis*, I gave a talk on simulation theory at Google. [3] Shortly thereafter, I met with an old colleague and fellow MIT alum, Bruce, whom I had worked with years earlier in Boston. He had just joined Google and was visiting the Googleplex in Silicon Valley. Not only is this odd collection of buildings the headquarters of one of the largest companies in the world, it is located in the commercial heart of Silicon Valley, just down the road from where I was living at the time in Mountain View, California.

Bruce, a sturdy fellow with thick glasses and a sharp but practical mind, and I were sitting outside a coffee shop on Castro St. This street, which sits in the center of Mountain View, a quaint little town at the bottom of San Francisco Bay, has a European flavor but with the added benefit of California sunshine. Since we were both computer scientists, we pretty much ignored the beautiful landscape of the Santa Clarita mountains to the west and the Fremont hills to the east that gave Silicon Valley its name, and immediately started to catch up and geek out.

Bruce had heard of my book, and we naturally started discussing implications of the physical world around us being some kind of simulated computer reality. Although we were initially talking about the kinds of computations that would be involved in generating and maintaining such an ultra-realistic simulation, at some point Bruce told me that he had been reading about the Mandela effect and that I should look into it further.

I had heard of the effect, which was about a subset of people remembering that Nelson Mandela had died in prison in the 1980s. Like most scientifically minded people, I had dismissed it as a fringe theory that could be easily explained away as faulty memory, since Mandela had actually died many years later.

Bruce then mentioned that the simulation hypothesis was actually the best explanation for how something like the Mandela effect could be occurring. This caught my attention, not least of all because he wasn't the kind of person I would've expected to bring up something as esoteric as the Mandela effect, let alone consider that it could be real or how it could be working. The people who brought up the Mandela effect to me were usually either discussing science fiction or were heavily into the paranormal world, bringing it up alongside topics often dismissed by mainstream science such as UFOs, ghosts, and Bigfoot.

I said I'd look into it. Bruce warned me that I had to be

careful, because the figurative rabbit hole in this case went pretty deep, and I was likely to be drawn in.

He was right. I started to explore case studies on various online forums about the Mandela effect. After digesting these, along with the various explanations from mainstream social scientists dismissing it as a case of mass faulty memory, I started to bring it up with some of my more open-minded scientist colleagues to figure out what it might tell us about time and space and simulations, particularly with respect to the idea of multiple timelines in quantum physics. They told me explicitly that if we are to take the findings of quantum mechanics seriously, then the past isn't what we think it is.

These discussions convinced me that if we were in a simulation, then multiple timelines were not such a crazy idea at all. In fact, it made some of the baffling findings for quantum physics that had been a key part of my argument in my previous book that we live in a simulation make *more* sense, not less. Multiple timelines in a simulated universe would actually be a *better* explanation for these mysteries than the worldview of a single, fixed timeline in a single physical universe.

Many of the confounding aspects of quantum physics are confounding only if we insist on a completely deterministic, materialist model of the universe, with a single past and a single future. The observer effect, the collapse of the probability wave, even parallel universes all make much more sense if the universe actually consists of information that is stored, processed, duplicated, and, most important, rendered as the physical world we see around us.

This book is an exploration of the possibility of a simulated multiverse, in which timelines other than what we experience as the main timeline might have existed (and might continue to exist). We will explore this complex idea through the lenses of science fiction, hard science, and good old speculation.

From our normal everyday experience, and from a classic

physics point of view, this idea seems like a logical impossibility. But if you think of it from the point of view of a simulated world, suddenly the idea of multiple timelines extending from multiple pasts into multiple futures doesn't seem so strange anymore.

The Strange Mind of Philip K. Dick

If the implications of all this sound to you like they might be more appropriate in a science fiction novel, particularly one by famed writer Philip K. Dick, then you and I are in the same boat. Dick was one of the most prolific and unique writers of science fiction in the twentieth century. In fact, my conversation with Bruce and later explorations into the topic brought me back again and again to my conversation with the late writer's wife, Tessa B. Dick.

I had asked to interview her because the Wachowskis, creators of *The Matrix*, claimed to have drawn inspiration from Philip K. Dick, and because I had heard of a quote from him that we were living in a computer-programmed reality. The quote was a famous clip from his speech in Metz, France, at a science fiction convention in 1977. Since he was one of the first in the modern era to talk publicly about this idea, I figured she could tell me what made him think we were living inside a virtual reality.

Dick's large body of work frequently explored two big questions: what it means to be human (versus nonhuman or, in the case of *Blade Runner*, an android), and how much of our experiences are actually real. That second question, about what is real and what isn't, had burrowed into my mind as I researched the simulation hypothesis.

Originally, like many consumers of pop culture, I had been familiar with Dick only through the various screen adaptations of his work. In addition to *Blade Runner*, some of my favorites included *Total Recall, Minority Report*, and the recent TV series, *The Man in the High Castle*, which was adapted from his 1960

Hugo Award-winning novel and was still going on when I interviewed Tessa.

One of the first things Tessa asked me was whether I had seen the whole Metz speech, not just the famous quote, which she repeated word for word:[4]

> We are living in a computer-programmed reality, and the only clue we have to it is when some variable is changed, and some alteration in our reality occurs.

I agreed to track down the whole speech if it was available online. My conversation with Tessa and subsequent readings of Philip's full speech, which was titled, "*If You Think This World Is Bad, You Should See Some of the Others,*" ended up being largely an interesting aside when I wrote *The Simulation Hypothesis*.[5] At the time, I was mostly interested in the first part of his statement, about being in a computer-programmed reality, a colorful way to get into the topic for fans of science fiction. I honestly didn't pay a lot of attention to the second part of the statement, or the rest of the speech, where Philip seemed to be saying even stranger things.

After speaking with Bruce and my initial research into the Mandela effect, I dove back into Dick's speech with gusto and dissected my previous interview with Tessa. This caused me to reassess what Dick had been saying from a wider, richer perspective.

I realized that Dick's ideas went much further than I had first thought and presented a very coherent, if somewhat speculative view, of how time and the universe work. The second part of that now famous quote, "...the only clue we have is when some variable is changed, and some alteration occurs in our reality," was perhaps the more important phrase that unlocked the rest of his thinking. It would not just mean we were living in a simulated reality, but that it could run multiple timelines. I realized that this is what the Metz speech was *really* about.

The Man in the High Castle and Alternate Timelines

In an eyebrow-raising moment during our interview, Tessa told me that Phil claimed to have remembered parallel timelines, which had a different history than the one we would call our consensus memory. According to her (and Dick himself, as I verified in the full speech), Philip claimed that his best-received novel, *The Man in the High Castle*, wasn't based solely on his imagination, but was based on *actual* "residual memories" of an alternate timeline.

Although it was always considered a gem in the world of science fiction, the general public is now more familiar with Dick's only Hugo Award-winning novel because of the Amazon adaptation in 2015. The novel takes place in an alternate timeline where the Axis powers, namely Nazi Germany and Imperial Japan, won World War II and have split the United States between them. Dick *claimed* that this was one of his residual memories of a brutal military state.

In a self-referential twist inside Dick's novel, *The Man in the High Castle*, a character, Hawthorne Abendsen, writes a book about an alternate timeline, one in which the Allies won the war and America was not divided between the Nazis and the Japanese. In essence, while Dick is giving us a glimpse of an alternate timeline, Abendsen's fictional book within the book, *The Grasshopper Lies Heavy*, gave the residents of that timeline a view of an alternate timeline—our timeline. The Amazon series ended up turning this literary device into a set of mysterious films that are newsreels from other timelines, which is an even more chilling experience, both for the characters and for the audience.

Although it's not unusual for science fiction writers to start thinking of their work as having taken on a life of its own, this was different.[6] Both Tessa and Philip were saying something more. Dick, in his Metz speech, admits that he had been obsessed with a dark version of events in America and that he actually remembered this timeline in fragments:

> Does any one of us remember in any dim fashion...
> nightmare dreams specifically about a world of
> enslavement and evil, of prisons and jailers and
> ubiquitous police?
> I have.
> I wrote out those dreams in novel after novel, story
> after story; to name two in which this prior ugly
> present obtained most clearly, I cite *The Man in
> the High Castle* ... and *Flow My Tears, the
> Policeman Said.*
> I am going to be very candid with you: I wrote both
> novels based on fragmentary residual memories of
> such a horrid slave state world...

Until 1974, Dick said he had only these "fragmentary residual memories." During that year, Dick claimed to have had a set of experiences which convinced him that he wasn't just writing made-up stories. During that time, he claims that *all the memories of the other timeline* came flooding back to him.

According to Dick, this was similar to what the Greeks called *anamnesis*, the return of memory from a prior life, though a more literal translation would be "loss of forgetfulness." The state of forgetfulness, according to the Greeks, was induced by crossing Lethe, the river of forgetfulness, when incarnating (i.e., being born). In the Metz speech, Dick continues to talk about the implications of this process:

> ...[T]he irony is this: that my own supposed
> imaginative work The Man in the High Castle is
> not fiction—or rather is fiction only now, thank
> God. But there was an alternate world, a previous
> present, in which that particular time track
> actualized—actualized and then was abolished due
> to intervention at some prior date ... I retain
> memories of that other world.

Dick also said that writing stories of an alternate world helped him deal with these dark residual memories. After his anamnesis, Dick said he no longer needed to write about these dark alternate timelines. Eventually, these memories faded "as would a dream upon the awakening of the dreamer."

An Alternative Previous Present and Glitches in the Matrix

What are we to make of Dick's ideas of a "previous present"? Should they be taken seriously, or are they just the ramblings of a highly imaginative mind?

Dick perhaps anticipated the incredulous reactions of many of the Metz attendees (clearly visible in the video clips) by including a disclaimer in the speech itself:

> You are free to believe me or free to disbelieve, but please take my word on it that I am not joking; this is a very serious, a matter of importance ... Often people claim to remember past lives; I claim to remember a different, very different present life. I know of no one who has ever made that claim before, but I rather suspect that my experience is not unique; what perhaps is unique is the fact that I am willing to talk about it.

How does Dick think these alternate timelines are formed? It goes back to the second part of his famous quote. According to Dick, it's all about changing variables and running the events again, which leads us to "relive" the same events again.

This idea that things have changed turned out to be part of his inspiration for his story, *The Adjustment Team*, in which the protagonist stumbles across a team of people who are responsible for adjusting reality. In the 2012 movie adaptation, called *The Adjustment Bureau* (starring Matt Damon and Emily Blunt), they are depicted somewhat like angels (though this was not indicated in Dick's original version of the story).

Tessa told me Phil wrote the story because of an incident when he went into the bathroom and remembered clear as day that the room had a light that could be turned on or off by pulling a chain. But the chain was no longer there; it had been replaced with a light switch. He wondered if someone or something was changing reality and his memory of the chain light was from a different version of the alternate present—a small detail that was one of many small changes resulting from an adjustment in a previous past that cascaded into the current present.[7]

The next few lines after the famous quote are also quite revealing, highlighting the central role of these little changes in his thinking:

> We would have the overwhelming impression that we were re-living the present—déjà vu—perhaps in precisely the same way: hearing the same words, saying the same words. I submit that these impressions are valid and significant, and I will even say this: such an impression is a clue, that in some past time-point, a variable was changed—re-programmed as it were—and that because of this, an alternative world branched off.

The idea of reliving a particular scene or experience but with variables changed was essential to the worldview he described in this speech. This idea that feelings of déjà vu were clues to the shifting nature of reality was strangely familiar to me.

In fact, his whole discussion had a weird sense of déjà vu for me personally. I had written a whole book, *Treasure Hunt: Follow Your Inner Clues to Find True Success*, about things that seemed off—feelings of déjà vu, synchronicity, or funny feelings, and I had used the same terminology, calling them "clues", perhaps to alternate possible selves in parallel timelines or future versions of us. I had even suggested that these clues were really "glitches in the matrix," a phrase that came from the 1999 blockbuster movie but is now commonly used for small, anomalous experiences that can't be explained.

Shifting Timelines and Programmers?

Dick's speech, if taken literally, presents many questions. If things were changing, who or what was changing them? Why are they being changed? And what happened to those old versions of the present? How do these alternate realities interact with our current timeline, if at all? In short, these are the subjects of the current book.

In what sounds like it could have come from inside one of his novels, Tessa went further and told me that Dick claimed that he was in communication with beings who told him that they had changed the timeline. They could watch the computer-programmed reality and then rewind it, change some variables, and move it forward again. This sounds eerily close to what we do when building and watching computer simulations, although the term *simulation* had not entered the popular lexicon at that time, and video games were in their infancy.

In fact, these beings were like Dick's fictional "adjustment team": supernatural beings, from our perspective, who could cause us to relive the present based on different variables and parameters. In simulation-speak, we might call these beings programmers or super-users who had the power to manipulate the simulation. In fact, Dick himself used the terms *Programmer* and *Counter-programmer* in his Metz speech, implying one or more beings that were changing the variables as if they were playing a game of chess with the universe we live in.

Tessa gave me another example of different timelines that Philip believed, which he didn't mention in his Metz speech: the assassination of JFK. According to Tessa, Phil told her these beings modified the timeline to try to prevent the assassination of JFK, not just in Dallas in 1963, but in other places. In some of these alternate timelines, JFK was assassinated in another location (in Orlando, for example), so their interventions were fruitless. In others, he wasn't assassinated, but that timeline went into a much worse place than our own (in some cases a nuclear war), so they reverted to our timeline.[8]

It seemed, in Dick's view at least, as if there was a particular reason for running these timelines: to make the outcome of the simulation *better* in some way.

Orthogonal versus Linear Time

Dick himself gives no definitive explanation of how all this worked, but he did have a high-level theory. He referred to the whole thing as "a lateral arrangement of worlds, a plurality of overlapping Earths along whose linking axis a person can somehow move."

In *The Man in the High Castle*, one character, the writer Abendsen, was sensing this other world in his writing, and another character, trade minister Tagomi, actually was able to go to an alternate timeline (i.e., our timeline). In the series, this alternate timeline was referred to by the Nazis as "Die Nebenwelt" or "the Other World" or "the World Besides."

Dick called his theory "orthogonal time." There was our ordinary linear time inside the computer-programmed reality, and then there was a perpendicular (or orthogonal) axis of time, which existed outside of the simulation, as it were. Tessa mentioned this concept when I interviewed her, but I hadn't really pursued it, so I had to go back to his speech and read his own explanations later.

> At no time did I have a theoretical or conscious explanation for my preoccupation with these pluriform pseudo worlds, but now I think I understand. What I was sensing was the manifold of partially actualized realities lying tangent to what evidently is the most actualized one, the one that the majority of us, by consensus gentium agree on.

What were these lateral worlds like?

Dick used the analogy of a closet with suits or shirts arranged next to each other. A person outside the closet could choose one, try it on, then try another, and that's what he said the programmers were doing. Why? In Dick's estimation, the tinkering was to make a better world, which is why the world we remember today is better than some of the worlds he remembers and writes about.

If this idea could only be dismissed as the drug-crazed rantings of a peculiar science fiction writer (which some have tried to do), then that would be that. But what led me to give Phillip K. Dick's ideas a central position in this book was that I found others, including respected physicists, who had their own versions of this idea, though expressed in a more formal way. This collection of ideas of multiple parallel timelines (or multiple parallel worlds, take your pick) is now informally called the *multiverse*.

The Multiverse Passes the Ten-Year-Old Test

Popular fiction often explodes with a new idea in parallel with the public's comfort level with new scientific discoveries. This was true with time travel and extrasolar planets in the twentieth century and is rapidly becoming true for the idea of the multiverse in the twenty-first century.

For example, the first modern time travel story was *The Time Machine* by H. G. Wells, written in 1895 during the thick of the Industrial Revolution. It was no accident then that Wells's time traveler was one of the first (if not *the* first) to use a machine to travel through time. Before the awareness of machines entered the public's consciousness, although there were some random examples of a character magically ending up out of time (think of Rip van Winkle or Mark Twain's *A Connecticut Yankee in King Arthur's Court*), there were no time machines.

In the twentieth century, a new type of story emerged that demonstrated this trend perfectly: the superhero story. To explain their superhuman powers, superheroes were often depicted as beings from other planets. Superman, the most famous example of this, came to Earth from the planet Krypton. This reflected a growing comfort level in the general public with our expanded knowledge of our solar system and galaxy. I like to say that Superman showed that exoplanets had passed the 10-year-old test: even 10-year-olds knew enough about the universe not to think it was odd that Superman came from another planet.

In the twenty-first century, the idea that a superhero comes from another planet is no longer novel; in fact, it's kind of old and boring. Today, it seems, creators of superhero stories are incorporating new ideas from science about the universe. As a result, we are meeting superheroes not just from other planets in our physical universe, but from Earth and other planets in other universes altogether, each of which has a different timeline.

Although the idea of parallel universes isn't completely new in science fiction (we'll be exploring many examples in the sidebars in most chapters), the idea really caught on in this century as evidenced by it passing the 10-year-old test with flying colors, thanks to the recent slate of TV shows about superheroes from DC Comics.

These shows, such as *The Flash*, *Arrow*, *Supergirl* (and others), all exist in a multiverse. How do I know this? This was explained to me quite matter-of-factly by my nephews (all of whom were under 10 years old at the time) in great detail as they pontificated about how time travel could create multiple timelines, and how the Superman of Earth 32 was different from the Superman of Earth 16. Since I was an "old guy" who might not get it, they felt it was their duty to lecture me about the multiverse, using their favorite superheroes to explain it in simple terms!

The Multiverse Graph and the Core Loop

The multiverse idea is so common now in the world of physics that physicists have proposed not just one but many types of multiverses. The one that was most interesting to me, given my previous research into simulation theory, was the many-worlds interpretation (MWI) of quantum physics, also known as the parallel universe theory, or the *quantum multiverse* for short. It is a well-respected explanation for the mystifying phenomenon of quantum indeterminacy by many physicists. In this interpretation, the universe is spinning off new branches every time a quantum measurement is made, resulting in an almost infinite number of parallel universes with some level of shared history.

If you laid this out in a graph, this would become the basis for one of the main models we will be developing in this book, what I like to call the Multiverse Graph. In short it is a map of the different possible states of the universe and the possible timelines between them.

Of course, as a computer scientist, the idea that we would be spinning off an infinite number of physical universes sounds somewhat absurd, because the number just keeps growing, as do the processing and memory requirements. The more logical explanation is that these universes are spun off and are stored as information and only loaded and rendered when they need to be, a process which I described in *The Simulation Hypothesis*.

Another way that computer scientists deal with infinite trees is to prune them along the way, cutting out unnecessary or undesirable paths that no longer serve the purpose of our computation. Basically, the infinite tree is pruned, as necessary by trying out different outcomes, and proceeding forward with only the most promising ones. We will delve into these ideas, along with quantum computation and quantum parallelism, in this book.

What this means is that the universe is not only a computing device, but that it creates tree-like structures, not just in space, but across time. Tree-like structures are present everywhere in nature, from the evolution of species to the evolution of languages. In computer science, tree-like structures are actually one of the most efficient and flexible ways to store any kind of data or set of nodes.

How are these tree-like structures across time created? Here we arrive at another central model we will explore in this book, which I like to call the Core Loop. In computer science, a loop is a piece of code that gets executed again and again. It might end up making different choices each time, and there are different ways to implement repetitive algorithms, but more or less, the same logic is executed each time until some boundary condition is met and the loop terminates, or it reverts to another loop running at a higher or lower level (called recursion).

In a sense, the central idea that I want to explore in this book is that the universe is a computer running the Core Loop, spinning off possible timelines, each of which is a path through the Multiverse Graph.

Where We Go from Here

As we've seen in this chapter, some of the ideas we are exploring in this book will sound like science fiction. Philip K. Dick reached some of these conclusions back in 1977, using his own terminology, and in a way, his ideas will serve as a blueprint for the topics that we'll explore in this book. Our tools will not just include science fiction references but will be simulation theory, information theory, video game architecture, quantum mechanics, and quantum computing.

In the end, we will come up not just with a model that can accommodate multiple presents and multiple futures but, also, multiple possible pasts. Like all good models, the models we explore in this book will turn the unexplainable into an inevitable conclusion, even explaining strange effects like the Mandela effect.

In the rest of *Part I: Sounds Like Science Fiction*, I explore ideas that are speculative enough that they sound like science fiction. From here, I will cover the Mandela effect in Chapter 2 and tie that into the simulation hypothesis in Chapter 3, which will also provide an overview of Nick Bostrom's simulation argument and a recap of my previous book.

In *Part II: Some Far Out Science*, we will switch gears to explore what science has to say about the ideas of time, space, and multiple parallel universes. We'll start with different versions of the multiverse idea that have been proposed by physicists in Chapter 4. Then we'll home in on quantum mechanics, because the version of the multiverse we are most interested in for our discussion is the quantum multiverse, which we will get into in Chapter 6. We'll then do a deep dive into what science has to say about time in Chapter 7, showing how in both relativity and quantum mechanics the past and the future are different from what we think of them as in everyday reality. Although this part contains the most science, we'll stay away from equations and keep it at the conceptual level, but this should be enough to convince you that time is very strange and not at all what it appears to be.

In *Part III: Building Digital Worlds*, we'll switch gears yet again to explore the idea of a digital multiverse. Using techniques and concepts from video games and classical computer science, I'll explore what simplified digital universes might look like. These techniques include creating a simple "gamestate" in a very simple adventure game, *SimWorld*, in Chapter 8. This will be followed by an exploration of cellular automata, which are very simple graphical computer programs that can exhibit complex behavior in Chapter 9. Then, in Chapter 10, we'll turn our attention away from simple computing structures to how the universe may actually be computing: using qubits and quantum parallelism, which provides a much richer platform for the multiple arrows of time. This is by far the most technical part of the book, including some light code (pseudocode) and logic gates, but you can get the core concepts in each chapter without going into the details.

In *Part IV: Algorithms for the Multiverse,* we will go back to the two core models introduced by this book to visualize how a simulated multiverse may be working. We'll do this by combining various concepts we have learned over the course of the book to represent multiple parallel universes as tree-like structures in time in *Chapter 11: Digital Timelines and Multiverse Graphs*, and how a computational process might navigate such a graph in *Chapter 12: The Core Loop as Search.*

Finally, in *Part V: The Big Picture*, we'll look at why we run simulations and how this relates to the core idea of this book, that we are constantly branching and merging universes. This will bring together the four main ideas in this book, parallel universes, the simulation hypothesis, quantum computing, and the Mandela effect in *Chapter 13: The Upshot—The Universe Evolves Through Multiple Simulations.* This will include a new metaphor based on an old story, "The Garden of Forking Paths," and will include discussions of other physicists who have proposed similar models, including the work of physicists like Thomas Campbell, Fred Alan Wolf, and Amit Goswami. We will end with *Chapter 14: Stepping Back—What Does It All Mean?*, where we'll zoom out and ask what, if anything, this might mean to us in our own lives. To do this, we need to move beyond science and technology and look at things from a completely different perspective: what is outside the multiverse? Could it be part of the spiritual dimension that so many religions have told us about?

If this all sounds to you like we've arrived somewhere on the boundary between science, science fiction and absurdity or, as the old show *The Twilight Zone* used to say, "the border between science and superstition," then you are in the right place. Welcome to life in a simulated multiverse.

TIME OUT OF JOINT AND A FALSE REALITY

Although Philip K. Dick often gives his characters reasons to question reality, one of the more pronounced examples of this comes from his novel, *Time Out of Joint*, which was published in 1959 and has not been directly adapted into film (though *The Truman Show* of 1998 had some similarities). The title of the book comes from a line in Shakespeare's *Hamlet* something is amiss in Denmark: *"The time is out of joint; O cursed spite! That ever I was born to set it right!"*

In Dick's novel, we meet Ragle Gumm, who lives in an alternate version of 1959. He lives in an idyllic little town with his sister and her husband. Gumm makes a living by solving a newspaper puzzle contest that appears every day called, "Where will the little green man land next?"

Strange things happen to Gumm in his quiet little community as the story progresses, which he at first thinks are hallucinations. He finds pages of a magazine with an article that features Marilyn Monroe, whom no one has heard of. He finds a soft-drink stand that has disappeared, replaced with a piece of paper that says "SOFT DRINK STAND," and he finds an old phone book with phone numbers from an exchange or town that doesn't exist. At one point, he finds a radio that was hidden away in someone's house and overhears military pilots talking, and they mention him by name. This seems more than a little odd. He even overhears a neighbor saying, "What if Ragle Gumm gets sane again?"

Gumm starts to investigate, even trying to leave the town, but fate seems to intervene, making it difficult for him each time. In the process, he discovers more oddities: a magazine with him on the cover, for example.

Eventually, Gumm escapes and learns what is really going on. The year is actually 1998 and the Earth is in the middle of a war with colonists from the moon, who are sending nuclear strikes to the Earth. The idyllic town he lives in is a construction set up just for him to look like 1959, a stable environment from his childhood. Ragle Gumm, it turns out, has a unique ability to predict where the next nuclear strike will land. Gumm, unable to shoulder this responsibility, went "insane" and retreated to his childhood. The ruse of the newspaper contest allowed him to continue to predict where the next strike will occur. The novel ends on a hopeful note; Gumm regains his sanity, leaves the artificial world of 1959 behind, and plans to leave Earth to explore the solar system.

Chapter 2

The Mandela Effect—Real or Mass Delusion?

What we would need at this point is to locate, to bring forth as evidence, someone who has managed somehow ... to retain memories of a different present, latent alternate world impressions, different in some significant way from this, the one that is at this stage actualized.

—Philip K. Dick, *Metz Speech (1977)*

Even Philip K. Dick recognized his idea of lateral worlds branching off in a computer-generated reality—his version of a simulated multiverse—would be very difficult to prove scientifically, since it was all based on his own memory. However, Dick recognized that if we could find other people who remembered different pasts or alternate presents, meaning they remembered a different timeline, this would lend more credence to his worldview.

It turns out that is exactly what the Mandela effect, which my colleague Bruce mentioned to me, was all about. It would have been difficult in pre-Internet times to corral a large number of people who might remember something differently. The Mandela effect has become so popular precisely because of the ability to collect memories from people in different locations instantly.

My goal in this book is not to try to prove or disprove the

Mandela effect; rather, it is to use it as a way to illustrate our larger idea, that of a simulated multiverse running multiple timelines, branching and merging in an ongoing computation. The Mandela effect, it turns out, is a colorful way to talk about the idea of multiple pasts that we can all relate to.

What Is the Mandela Effect?

The name of the Mandela effect refers to people remembering Nelson Mandela's tragic death in prison in the 1980s, even though he actually survived until long after his release. The term itself was coined by blogger Fiona Broome, when she encountered a number of people who remembered things that hadn't happened, ranging from Mandela's death in prison to episodes of *Star Trek* that don't seem to exist. On her website, Broome has collected many examples of this effect, where a large number of people remember things differently than they actually happened.[9]

Broome herself notes that in the decade since she coined the term, the phrase has gone mainstream. It was even referenced in the recent reincarnation of *The X-Files* and has become a popular Internet meme[10]. Many mainstream news sites have had articles about it, going over examples of these effects[11], and recognizing its popularity with the public, many scientific news sites have published articles debunking it.[12]

Let's start with a slightly more formal definition:

> **The Mandela effect** is a phenomenon in which the minority of the public retains memories of past events (or objects) that are different from what the majority remembers.

You'll see that when we talk about a single Mandela effect (an "effect" or "effects"), we are talking about (usually) two versions of a single *item* (event, picture, object, quote, etc.): what the majority remembers and what the minority remembers. What the majority remembers, and what is documented online and by historians, would be what Dick would refer to as the "consensus gentium" present reality. Sometimes, there is more than one alternate version of the item in question, which could be defined as multiple effects taking place at different times.

At least to the minority of those who remember something different, the current present reality implies that somehow, somewhere along the way, either in the past or the present, an alteration was made. Alternatively, you could say the alteration was made in the memory (of either the majority or the minority) to conform with the new present timeline.

In this model, the majority has no memory or history of the alteration despite anecdotal remembrances from people in the minority. In some cases, there is some circumstantial evidence to support the minority's claim that something was different, but nothing definitive.

There is, of course, a conventional explanation of the Mandela effect that has nothing to do with timelines: that it is simply faulty memory. We'll talk about this and other possible explanations later in this chapter.

Since Broome's first naming of the Mandela effect, many examples of effects have spread, primarily online in forums on Reddit and social media. One particular Reddit forum (r/mandelaeffect) is a treasure trove of people posting possible effects and others commenting on whether they remember the *alternate* or the *current* version of events.

As my friend Bruce warned me, online forums like that one can become veritable rabbit holes, containing not just the popular well-known effects but also many obscure effects that may have been experienced by only a few people. For our purposes, since we are concerned with merging timelines and diverging timelines more than individual variances, we'll focus only on those effects that have been reported by many people.

Categories of Mandela Effects

It's easier to understand the scope of the Mandela effect and to come up with explanations for individual effects by taking a bird's-eye view of the landscape. I believe this is best done not by listing 30 or 40 of them individually, as many online articles do, but to break them into categories.

My rather informal categorization of the most interesting effects includes the following buckets:

- Major Events/Deaths
- Film/TV-Related
- Spellings and Logos
- Geography
- Religious Scripture
- Physical Objects

Before we go into possible explanations for the Mandela effect or how it relates to overall thesis, let's go through a few examples in each category.

Major Events/Deaths

The most startling examples are ones that involve nonexistent events that many people remember happening, such as Mandela's supposed death, after which the effect is named.

Broome writes on her website that she thought Mandela died in prison:

> I thought I remembered it clearly, complete with news clips of his funeral, the mourning in South Africa, some rioting in cities, and the heartfelt speech by his widow.
> Then, I found out he was still alive.[13]

In our timeline at least, after spending almost three decades in prison, Mandela was released in 1990. It was a global news event that was "… watched by millions across the world."[14] The next year, in 1991, he became president of South Africa and shared the Nobel

Peace Prize with Frederik Willem de Klerk, the last white president of the country who oversaw the end of apartheid and transfer of power.

Broome then interviewed hundreds of people and found they also shared the alternate memory of Mandela dying in prison, complete with details of the funeral along with details of where they were when they heard the news.

At the time the name was coined, Mandela was still alive, so it couldn't have been his funeral that they were remembering, could it? One explanation often given is that they were remembering the death of another South African in prison, Steve Biko, who died in 1977. But this ignores that when people remember where they were upon hearing of the death or watching the funeral, it was during the 1980s (and in a subset of cases, it happened even later, in the 1990s).

This explanation (faulty memory) and many like it for the effects to follow could perhaps be accepted easily (as many mainstream scientists do) but for the factors of *significance* and/or *proximity*. Unlike, for example, a misspelling of a brand of peanut butter, for some people Nelson Mandela had a particular significance. In some of the effects we'll explore, there was both significance and proximity to the subject.

For example, YouTube blogger Eileen Colts tells how she, as a journalism student, actually went to South Africa to try to interview Mandela in prison, but she couldn't because she was told he was "very ill." Later, when she graduated and was working at an NPR station in Chicago, she remembers: "In 1986 or 1987, I specifically recall hearing reports at work that were broadcast of Nelson Mandela's death in prison, tragically just weeks before his release was finalized."[15] She goes on to say that she also recalls his widow, Winnie, taking over as the leader of the resistance movement, a detail that is echoed by others who remember this timeline. Stories like Eileen's are not that uncommon and are usually ignored by those who, to borrow an *X-Files* term, "want to believe" that the all Mandela effects can be casually dismissed.

In some cases, it's not the event itself, but the date that is remembered differently. Most of us know that the space shuttle *Challenger* exploded in 1986, but many insist it was in 1985 or even as far back as 1983, often citing specific things they were doing or which class they were in when it happened.

Going further back, those who were alive during the Lindbergh baby kidnapping of the 1930s know that Charles Lindbergh's baby was found in 1932. It was national news during a time when the news media wasn't so varied as it is today. However, some subset of people claimed to remember that the baby was never found. Similarly, many people claim to remember watching documentaries and wondering what had happened to the baby that was never found.[16]

Similarly, some claim to remember the preacher Billy Graham dying sometime between the 1990s and the early twenty-teens, and even remember details such as Bill Clinton speaking at his funeral. However, in our timeline, Graham died in 2018. Being one of the more recent effects, this was one that surfaced on Reddit many times. In 2015, an entry reads:

> I distinctly recall, a couple of years ago (... 2013 or so) my grandparents (who I live with) telling me that Billy Graham had passed. They are both staunch Christians who followed Graham's ministry for a very long time, so I took this at face value. A few days later, an evangelical magazine arrived in the mail (I remember because I fetched the mail that day) with a cover story on Graham's death. His picture on the front and everything. A short time later, the grandparents went out of state to attend some big Christian conference. When they returned, they told me how someone (possibly Graham's son? not sure) got up and gave a lovely speech, some nice words about Graham's life and memory. They continued to talk about the man's death for some time after.[17]

There are many who remember Graham's funeral, recalling specific details. Personally, not having paid attention to Graham or Christian ministers, I wouldn't have the slightest idea whether one of them passed away, let alone the year it had happened. You might say it carried very low significance for me. On the other hand, those who followed Graham's ministry religiously, just like those of us who follow, say, one of the lead actors in Star Wars (Harrison Ford or Mark Hamill), would be keenly aware of when news of their death was received. Similarly, you might say that Mandela's death had high personal significance and high proximity for Colts, the journalist who went to South Africa to interview him and was unable to.

In a twist on what we now know happened, many people remember seeing a young man run over by a tank in Tiananmen Square in Beijing in 1989. A quick search reveals that the young man was blocking the tank's path, but it did not run him over. From Broome's own website, some remembrances of this include: [18]

> I remember "tank boy" getting run over by the tank at Tiananmen Square. My husband doesn't. We googled it and apparently he didn't get run over. I have a very vivid memory though. I remember seeing a video of it. I remember learning this in 7th grade history.
>
> —Angel, September 2011

> I remember TANK BOY getting run over. My partner and myself were talking about Tiananmen Square and tank boy. I mentioned how horrible it was that he was killed, my partner had no memory of that and thought I was crazy. He had to go on YouTube to show me that he lived. As I watched I had no recollection of that event of him living.
>
> —Bree, August 2012

> Same here I remember seeing blood on the street after the tank rolled over him and how the backlash

nearly caused communism to fall apart in china [sic] and then they switched to the capitalistic command economy. This is so weird.

—*James, Date Unknown*

What's going on here with this memory of events unfolding differently than as Philip K. Dick would call it, the consensus gentium memory?

We'll explore various possible explanations later in the chapter.

Film/TV-Related

Both movies and TV are parts of our modern culture that have become fairly universal, often replacing the myths that were cultural touchstones of old. Several subtypes of the Mandela effect are related to movies and TV, and these are among the most popular examples, though perhaps relatively easy to write off.

The simplest examples are the misremembering of lines; the most famous, was Darth Vader's line in *The Empire Strikes Back,* which many misremember as, "Luke, I am your father." If you watch the film again, you'll see Vader actually said, "No, I am your father."

But Vader is by no means alone. As far as lines go, many of us remember the evil queen in *Snow White and the Seven Dwarfs* asking, "Mirror, mirror on the wall," but what she actually says was, "Magic mirror on the wall."

Misremembering of titles of TV shows is also common. Many people remember *Sex in the City* as the title of HBO's hit show starring Sarah Jessica Parker from the 2000s. A quick Internet search reveals that it's actually *Sex and the City*, which was the title of the book that the show was based on.

Moving on from simple cases of a word change, we find scenes that play differently than remembered, or episodes of TV shows and entire movies that people remember distinctly but which do not exist. As an example of a scene that is different, in the movie *Risky Business*, many remember Tom Cruise dancing around in his underwear, wearing sunglasses. In fact, he wore sunglasses in the movie, but not in that scene.

Broome herself writes of an example at Dragon Con, an annual comic book type convention held in Atlanta. There was an episode of *Star Trek* (the original series) that many fans present insisted they remembered. However, the cast members of the original series who were there at the conference insist this was never filmed. Those who have attended sci-fi or comic conventions know there are some die-hard fans of these shows who not only remember more details of shows than some of the stars but can quote lines from them which makes this case more puzzling.

In one famous case, there is a whole film that never existed (or at the least is being confused with another film). Many people claim to remember a movie from the 1990s, starring the comedian Sinbad, called *Shazaam*, in which a kid who wishes to a genie that his father would find love again.

> Meredith Upton, a 25-year-old videographer from Nashville, Tennessee, also remembers the same film. "Whenever I would see Sinbad anywhere in the media I would recall him playing a genie," she says. "I remember the name of the film as Shazaam. I remember two children accidentally summoning a genie ... and they try and wish for their dad to fall in love again after their mother's passing, and Sinbad can't [grant the wish]."[19]

Meredith isn't alone; there are possibly hundreds or thousands of people online who remember this movie. Some remember owning the VHS tape and distinctly remember forwarding and rewinding to specific scenes that they wanted to watch again. They also swear that the genie movie with Sinbad was different from the "other movie about a genie starring a black superstar" from the 1990s. That movie was actually called *Kazaam*, which starred NBA superstar Shaq (Shaquille O'Neil). At least in *our timeline*.

This particular effect grew so popular online that it prompted

Sinbad to tweet that apparently, he was in a movie he had no memory of. In a strange bit of fan service, in 2017, Sinbad actually shot a short scene of himself playing a genie that matched the remembrances of those online of two kids finding a lamp and a genie emerging to grant them a wish.[20]

Spelling and Logos

Perhaps the largest category of effects includes individual misspellings. This category might also be the easiest to explain away as faulty memory.

The most famous example includes the Berenstein Bears (which is actually spelled Berenstain Bears), which we'll talk more about in a minute. But there are many other effects like this. Almost as well known is Jif peanut butter, which many people remember (including me) as being Jiffy peanut butter, but there is no Jiffy, only Jif.

Another popular one is about the "Looney Toons" cartoons, which feature iconic characters such as Bugs Bunny. In fact, there is no Looney Toons—it's called *Looney Tunes*. Yet another one is Oscar Meyer wieners, which is actually spelled Oscar Mayer, with an *a* rather than an *e* after the *M*. In a reversal of the Looney Toons example, the sugar-packed cereal Froot Loops, which I ate often as a kid, is actually Fruit Loops (the first word is spelled properly). Staying on the cartoon track, *The Flintstones* is remembered by many as "The Flinstones" (without the middle *t*).

Similarly, there are slight changes to fictional characters and logos that are different than people swear they remember. As one example, the Monopoly guy, featured in the famous board game, does not have a monocle, yet many people swear he does. Are they simply confabulating him with Mr. Peanut, another colorful character, who does in fact wear a monocle? Speaking of cartoons, there is the monkey, Curious George. Many people remember him with a tail, but if you do a quick search, you'll realize there is no tail.

There was, according to our timeline, never a tail on the curious little monkey. Similarly, for a younger generation, Pikachu's tail doesn't have a black part at the end; it's just yellow.

Blending logos with images: many of us remember KitKat as being spelled "Kit-Kat" with a dash, but that's not the case. Similarly, many remember the Ford logo without the little flair at the end of the middle part of the F, but there it is. According to online sources, it has always been that way since the early 1900s, when Henry Ford started the Ford Motor Company.

Investigation of this category of effects is complicated because most of the memories are from childhood, and they involve very small details, which makes most laypersons' memories perhaps more suspect than if the memory had occurred when they were adults.

However, even within this category, there are instances when individuals were either more knowledgeable or had more at stake (i.e., higher significance and/or proximity) who remember it differently, making it less likely that they made an obvious memory error. For example, there are cases of Jewish kids wondering why the "Berenstein" Bears were Jewish, reflecting the Jewish spelling "stein" and not "stain," which would not imply a Jewish background. Some remember having conversations with their adult relatives about the "Jewish bears." Surely the adults would have pointed out the spelling mistake if that's all it was?

Geography

A number of effects relate to memory of the location of specific places on the maps. There don't seem to be as many of these, so I won't go into many details. Two common ones are the relative positions of landmasses to Australia (whether New Zealand is located to the east, the northeast, southeast, or even to the west) and South America (whether it is directly under North America or out into the Atlantic).

Religious Scripture

Perhaps more troubling and more difficult to dismiss is the category of changes to the Bible or other religious scripture. There is a whole subculture now that is devoted to seeing changes in the Bible from the time they learned the verses as children. Countless people who say they spent many hours memorizing certain phrases word for word are waking up one day to find that the phrasing of the verse in their Bible is now different.

Now, the easy explanation is that they used a different translation when they were younger, or perhaps a new edition was released. After all, the King James Bible, the most popular Bible in the West, was translated into English from Latin, and there are bound to be different translations. But these people sometimes insist that it's not only changed in a new version, but that their physical copy of the Bible, which they have kept since they were children and still possess, is now different from what they memorized!

Now if it was simply a random phrase from a random book, that would be one thing, but great effort is made by religious people and preachers to remember passages from the Bible exactly as they are written (or at least as they are translated into your language). Like the memories of the reverend Billy Graham, they are less likely to have gotten the wording so wrong.

One of the most famous examples of this is the famous phrase, "The lion will lie with the lamb." Not only is it one that many Christians (and even non-Christians like me) remember, but there are even paintings with a lion and a lamb together reference, right on the picture, Isaiah 11:6, the verse from which this phrase is taken.

Now, the King James Bible says: "The wolf also shall dwell with the lamb, and the leopard shall lie down with the kid; and the calf and the young lion and the fatling together..."

Lest you think it's an issue with the one translation, there are other versions, such as the New American Standalone Bible, that say something similar: "And the wolf will dwell with the lamb ..."[21]

Another famous one is the Lord's Prayer, which even non-Christians have heard; it includes the phrase, "and forgive our trespasses." But if you look at Matthew 6:12, it actually says: "And forgive us our debts, as we also have forgiven our debtors." This one has perhaps been explained by the translations of a specific word that is use used by different denominations of Protestants.[22]

Nevertheless, these are just the most well-known examples; there are many less prominent examples that are equally, if not more baffling.

There are people who believe someone, or some force, is actually changing the Bible verses by messing with our reality, and there are many websites dedicated to pointing this out.

Has the Bible actually changed since it was memorized by these religious people? While some write off the whole thing as translation issues, others believe that there is a satanic force at work here. We'll discuss this theory as well when as we get into possible explanations.

Physical Objects

While misremembering the *t* in shows like *The Flintstones* (another common effect) is probably not cause for particular concern, or likely to result in a change in our understanding of the universe, when thousands of people remember specific details of events that we can safely say never happened, the conversation becomes more interesting. When there is physical but circumstantial evidence of the way things "used to be it becomes even more puzzling.

In the last section on Bible verses, one of the things that's quite surprising is that you can find online (and if you search your own memory) examples of physical objects that portray the lion and the lamb and often include a quote from Isaiah on the physical picture. This constitutes at least circumstantial evidence that there was a version of the phrase that said "lion" and not "wolf," which indicates that there was a change, whether the explanation for it is mundane or supernatural or scientific.

An area that I stumbled across as I researched this was well-known works of art that seem to have changed their posture or pose. One of these is the *Mona Lisa*, whose smile many claim has changed.

Perhaps more interesting is the speculation around *The Thinker*, the famous statue from Rodin.[23] In it, the figure has his hand just under his chin, with fingers pointing at the throat. Now, a popular thing for tourists to do when standing next to a statue is to try to re-create the pose of the statue, which millions of tourists have done all over the world.

Why, then, are there tons of pictures of tourists and well-known figures reproducing the pose of *The Thinker*, but instead of having their hand under their chin, they have their hand on their forehead?

And it's not just individual tourists. George Bernard Shaw, who often posed for Rodin, posed for a picture the night that *The Thinker* was unveiled in London. If you look at the picture by Coburn, its caption is even, "George Bernard Shaw in the Pose of the Thinker." Yet, in this photograph, Shaw clearly has his hand on his forehead.[24]

Adding to the weirdness, Rodin himself wrote about *The Thinker* and described it as having a clenched fist, which the current version of the statue doesn't seem to have. You might think a sculptor like Rodin would be aware of the difference. It's one thing to make a mistake of remembering some detail of a cartoon from fifty years ago differently; it's another to take a picture right next to a statue and get the pose completely wrong!

Now, there is a prosaic possible explanation. For one thing, Rodin originally designed *The Thinker* as part of *The Gates of Hell*, which is based on Dante's *Inferno* (and the figure was called "The

Poet"). The Poet sits at the top of the gates looking down on everyone. Rodin decided to enlarge it and make it into a separate statue, and it became one of his better-known works (if not the best known one).

Figure 1: Picture of George Bernard Show in the pose of The Thinker *(Colburn, 1906)[25] and picture of* The Thinker *from Rodin[26]. Something is clearly off.*

If you visit a Rodin exhibit at a museum like Stanford's Cantor Arts Center (which houses bronze casts of both *The Thinker* and *The Gates of Hell*), you will realize that Rodin went through a process that started with plaster originals. He then cast it in stone or, more commonly, metal (typically bronze). And before plaster, in some cases he must have started with sketches. With *The Gates of Hell*, the original plaster was restored and is available in Paris, from which the numerous bronze casts were made, the first of which was made in the 1920s (after Rodin's death) and is currently at the entrance to the Rodin Museum in Philadelphia.

So, is it possible that the original version he envisioned had *The*

Thinker with his fist on his head, either as part of *The Gates of Hell* or as a separate statue? And when G. B. Shaw posed, he was posing without the actual first bronze cast? This is certainly possible. However, the first bronze casts appeared in 1904, and the G. B. Shaw picture was from 1906, the night of an unveiling of one of the bronze casts in London. Moreover, this really doesn't explain all the pictures of people next to the bronze casts in the museums with their hand on their forehead from recent years.

Here we see both significance and proximity, and even circumstantial physical evidence. Although I haven't seen many of these types of effects, it doesn't mean they aren't out there, making it one of the weirder aspects of the Mandela effect, whether you believe it is due to multiple timelines or has a simpler explanation.

Possible Causes of the Mandela Effect

Although there has been quite a bit of research on how we form memories and store them in the brain, we still don't fully understand how memory works. This means that any theory which attempts to explain the Mandela effect must remain only that: a theory. Let's take a look at the prevailing theories.

It turns out that upon examination, although many commentators have done so, I don't think you can make sweeping generalizations about the origins and meaning (if any) of the Mandela effect. The reason I like to divide the examples into categories is that it's possible different categories might lend themselves more easily to a particular theory or explanation.

These explanations start with simple ones related to memories and then go to more speculative explanations, like religious or conspiracy theories, and end with more scientific explanations, culminating with the simulated multiverse idea. In this final explanation, the Mandela effect becomes not an anomaly but a built-in feature of how a digital, simulated multiverse works.

Simple Errors in Memory or Perception

My first reaction to the simplest examples, including wording and spelling related effects, such as in the Berenstain Bears case, was that these are most likely the result of simple errors of memory.

There's a common exercise in which you ask people to read this sentence and tell you what it says (about Paris in the springtime).

> I LOVE
> PARIS IN THE
> THE SPRINGTIME

Initially, people usually don't notice the extra word in the sentence, unless you ask them to point to each word as it's being said.

When we're reading, we often skip words or allow our minds to fill them in A 2011 study in the *Journal of Experimental Psychology: Human Perception and Performance* showed that words that are skipped are filled in 8% to 30% of the time.[27] An important factor in this study was the predictability and the length of the word: Since the repeated word, in this case, is very short and common ("the"), it is often overlooked.

What about the letters inside words? As Arizona State University associate professor Gene Brewer, PhD, explained to *Mental Floss:* "When you recall an event, you use memories around it, taking elements or pieces of other events and fitting them where they make sense."[28]

In other words, a likely explanation for the misspelling of Berenstain and Jif is because kids may have misspelled it, and even newspaper articles and school publications may have misspelled it. Similarly, this might explain small changes in logos and other effects when not much significance is attached to them. Nevertheless, the fact that a large number of people who have the same memory is what makes it a valid and interesting effect. Moreover, this simple explanation simply ignores the data of those who remember an effect with high significance and/or proximity.

Intentional False Memories

This is where we start to get more speculative on the reasons for and causes of the Mandela effect. Psychologists have found that telling people about a false event can cause them to remember it as true. This has happened in court cases and in test experiments. University of Virginia professor of psychology Jim Coan created the "lost in the mall" procedure[29] when he was an undergraduate at the University of Washington, Coan described childhood events to his family members, including one about his brother getting lost in the mall. His brother later took that to be a true event, even though it wasn't. The technique was applied to a larger number of people by his professor, psychologist Elizabeth Loftus, who found up to 25% of the participants remembered something that was false. Since then, this explanation has been used in court to "discredit abuse survivors' testimony by inferring that false memories for childhood abuse can be implanted by psychotherapists."[30]

Although there is validity to the idea that false memories can be implanted this way (effectively by suggestion by some external, malicious agent), this would mean someone was specifically *trying* to plant these false stories so that a large number of people might remember them as true. In the case of Mandela's death (and many other effects), this doesn't sound like a reasonable explanation. Unless there were multiple news outlets, including newspapers, TV, and radio, that were all in on it, either reporting his death in the 1980s or remembering it later), this explanation just doesn't hold up in my opinion. In fact, this explanation (that these are false memories intentionally planted) starts to sound like a conspiracy theory.

Are there other ways that false memories can come together? The Deese-Roediger-McDermott procedure allows for closely related words (like *bed* and *pillow*) to suggest other words (like *sleep*) that weren't in the list, but which participants remember as having been on the list. [31] This could be due to a failure of memory and how it works through association. But again, whereas this explanation seems to hold true for simple errors, it is unlikely that it could be responsible for major events, in which people remember Bill Clinton and other presidents honoring Billy Graham at his funeral, and some remember discussing it with relatives well before his actual death.

A more science fiction explanation would be that some super-psychiatrist outside of our normal purview was implanting false memories into the brains of a subset of the population as part of some experiment. Now we are back in the realm of Philip K. Dick's novels, in which false memories feature prominently. These ideas were carried forward in movie adaptations of Dick's work like *Blade Runner*, in which the android Rachel is given false memories of a childhood she never had, and *Total Recall*, in which Arnold Schwarzenegger's character is able to take a vacation by simply implanting memories in his mind. In both cases, the memories were planted by an external agent for a specific purpose.

What purpose would there be for implanting false memories in a large percentage of the population? We can only speculate about both purpose and the method of such a procedure.

Bad Actors: Satan, Jinns, and the Counter-Programmer?

As for changes in scriptures, an explanation often given by religious figures is who notice these changes is that there is someone or some entity causing these changes *on purpose*. This someone is outside of time and space and can influence both time and space—meaning some kind of divine or demonic being. Most often it is considered some kind of satanic entity that is messing with God's word.

Interestingly, it's not just hardcore Christian websites that believe the Mandela effect is real and are saying there might be other entities that are somehow interfering with the timeline and the word of God.

Although he didn't talk about the Bible, it turns out this explanation of someone making changes is closer to what Dick believed than I first gave it credit for. Dick explains that there is a Programmer and a Counter-Programmer who are each making changes opposite each other, as if they were sitting at a chess board to see what happens each time they change a variable. These changes can occur in the present or what we call the past, but since these entities live outside of linear time, we can see the repercussions across the timeline even if the change was made to the past.

A Muslim cleric, Shaykh As-Sayed Nurjan Mirahmadi, a founding trustee of the Islamic Supreme Council of America and president of one of the Sufi orders, makes the point that there are entities (known as *jinns*, usually in Western translations called genies) that exist outside of our normal ideas of space and time. In the Islamic tradition, these jinns have been given freedom by God (Allah) and are able to make changes to our timeline.

In Islam, there is a type of holy man called a hafiz, who memorizes the entire Koran word for word. Mirahmadi says that the reason for this is so that even if the written word of the Koran changes, the exact wording of the Koran can be preserved through the memory of all the hafizes.[32] In a way, this is an old version of the modern idea of data duplication to prevent tampering, and presumes that the jinn cannot interfere with human memory.

This presents the question: Why would someone intentionally change the holy scriptures of one or more major religion? And who would want it changed? Once again, we find ourselves in the realm of supernatural beings who might want to muck with the timeline from outside the timeline and we can only speculate as to possible motives.

Multiple Timelines and Time Travelers

Let's move away from supernatural beings causing timeline changes, to a popular potential explanation for the Mandela effect that is more relevant to the subject of this book: that there are multiple timelines.

Note that an explanation of more than one timeline by itself doesn't require a full multiverse, just two timelines with different parameters and small changes. As shown in Figure 2, there would be timeline A and a timeline B. If Timeline B is the consensus timeline, those who were part of timeline A would remember events differently from the rest of the world.

Timeline A + B

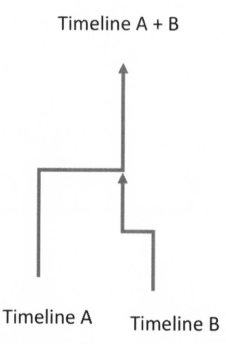

Timeline A   Timeline B

Figure 2: Example of merging timelines

If this sounds like we are back in the realm of science fiction, you'll recall that Philip K. Dick believed some type of timeline manipulation was going on by altering variables, as I described in the previous chapter. This could include small changes, like a light switch that was different, or different memories because the timeline was adjusted, or major events like the end of World War II or the prevention of the assassination of JFK.[33]

When you consider the sheer number of Mandela effects, though, a simple two-timeline explanation may not be enough to accommodate them all, since some people might remember each effect differently, resulting in a combinatorial number of timelines – i.e., a multiverse.

Quantum Explanations: Parallel Universes, the Multiverse, and the Possible Pasts

In this book, the point is to explore a more scientific framework, one that could explain the Mandela effect and some of the strangeness of quantum physics and redefine our understanding of time and space.

Quantum mechanics has an interpretation that allows for multiple timelines across multiple parallel universes, with a new universe branching out each time a decision is made. This would allow for universes where each combination of the Mandela effect happens or doesn't happen, in combination with each other effect.

In the multiverse case, just like in *The Arrowverse*, all possibilities already exist as separate universes. We just happen to be in the one where Mandela was alive until 2013 while the other ones still exist somewhere in the multiverse. There could be a universe where Mandela is alive, but the Bible is different, and on and on. But what could explain two people being in this timeline who have different memories of the past?

I would argue that the weirdness of quantum mechanics doesn't preclude multiple possible pasts, just as it doesn't preclude multiple possible presents or multiple possible futures. We'll talk more about this in Chapter 7, where we'll discuss the delayed-choice experiment. The odd results of this experiment tell us that there are multiple possible pasts and one is selected when a measurement is made. If we combine the delayed choice with the many-worlds interpretation, we reach the conclusion that there are multiple pasts, just as there are multiple presents and multiple futures.

Theoretically, if the multiverse does branch in this way, could the Mandela effect simply be about remembering one of the many universes? This is one of the themes we will be exploring in detail in this book in Parts II and III.

CERN and Tunneling to a Parallel Universe

One oft-tossed-out speculative explanation in the online jungle of Mandela effect chatter is that the effect was somehow caused by experiments at CERN, which altered the timeline or, rather, made touch with an alternate timeline or parallel universe.

CERN, located near Geneva, is the home of the Large Hadron Collider, the largest particle accelerator in the world, in which particles are accelerated to incredible speeds and, in some cases, smashed together. Many experiments have been carried out since it went live in 2008, resulting in important discoveries, including that of the Higgs boson particle in 2021.[34]

Does CERN do research into parallel universes? CERN itself has a research group that definitely talks about them as being "concrete physics theories that scientists are trying to confirm"[35] In fact, they hope to be able to find higher dimensions and then relate them back to our standard four dimensions, claiming that "parallel universes could also be hidden in these dimensions."[36]

The online chatter suggesting that CERN might have been responsible for the Mandela effect became so popular that in a statement to CNBC in 2017, CERN scientists had to deny that they had stumbled upon an alternate timeline (one that supposedly included the election of President Donald Trump).[37]

Could it be, however, despite their denials, that these atom smashers are somehow blasting a hole or doorway to another universe? Within physics circles, one of the ways of creating such a doorway to another, parallel dimension is thought to be black holes. We will delve a little bit more into the science of this in *Chapter 4: A Variety of Multiverses*. CERN specifically tells us that it may someday be able to create tiny black holes, called quantum black holes, but "scientists are not at all sure that quantum black holes exist."[38]

Whereas CERN generally rules out the possibility that it has contacted other parallel universes, it doesn't rule out the possibility of detecting extra dimensions of reality by creating tiny black holes in the future. CERN clearly tells us they haven't come even close to being able to do this yet, and most physicists would agree that this isn't a plausible explanation, at least in the present.

The Digital Simulated Multiverse

We now come to what is my favorite explanation for the Mandela effect, and the reason I went into so much detail about the effect: that we are living in a computer-generated reality, or a simulation, like a video game. Thus, as in any game, the parameters can be changed, and any scenario can be run multiple times. Each time a variable changes, it results in a number of ripples to the timeline, both small and large.

The simulations that we run on computers today are primarily deterministic processes —that is, processes that don't necessarily have random values or free choice, yet they still require simulations to run through a certain number of steps to get to the eventual outcome that resulted from any change, no matter how small. This is the basic idea behind chaos theory, in which small changes in parameters can lead to large changes down the road (called sensitivity to initial conditions, which we'll talk about in *Chapter 9: Simulation, Automata, and Chaos.*

> **Simulated multiverse**: a universe that is simulated on a digital computing system of some kind and can spin off different instances of the universe by branching, run them for some time, and then harvest results and merge them back into a main universe.

We will spend a lot of time in Part III and Part IV, showing how toy universes inside computers are created, maintained, saved, and evaluated, using video games and other software techniques.

Should the Mandela Effect Be Dismissed Outright?

If one person reports that they saw a giant sea creature, it is reasonable to dismiss the claim perhaps as a form of misidentification. However, when the number of people who have seen certain phenomena rises, it behooves us at least to look more seriously at what might be happening, rather than simply dismissing it outright.

It's of no use to say you don't believe in the Mandela effect. The effect isn't really a matter of belief; it is real effect in the sense that people report it. Different people have different memories of sometimes inconsequential details and sometimes consequential ones, including major events. When the number of people who misremember a specific event becomes large enough, we should start to think of this phenomenon in the same way as we think of something like other mass-sighting phenomena.

Science generally has a problem with mass sightings, such as the Phoenix Lights UFO sighting in 1997 or the Miracle of Fátima in 1917, both of which were seen by literally *thousands* of people. These sightings, which are hard to explain by any open-minded investigator, are easily dismissed by scientists.

Unfortunately, this is done by simply throwing away the inconvenient data and only concentrating on data that fits whatever theory the majority of scientists have agreed on. But if you look closely, you'll see that science as it's framed today doesn't have the tools to really investigate mass sightings or other phenomena that are ephemeral and sporadic, being difficult to reproduce on demand in a laboratory setting.

Most serious mainstream discussions of the Mandela effect include psychologists who dismiss the effect as simply a case of faulty memory. I believe the Mandela effect, if taken seriously, could also reveal something very interesting about the nature of time and the relationship of the past to the future. As I said at the start of this chapter, I'm not here to convince you of the reality of the effect or the timelines implied by the effects, but to use this a jumping off point for our exploration.

Some of the effects may in fact just be cases of faulty memory. There may also be other prosaic explanations. But the reason I chose to do such a deep dive into the effect in this chapter is that it provides a very powerful framework for thinking about the idea of multiple pasts, which is a component of a simulated multiverse model that includes multiple timelines branching and merging.

Just as Einstein's theory of relativity only deviates from Newtonian mechanics at the extremes, yet still reveals new properties of how the universe works, the Mandela effect might provide an extreme case of memory. It may reveal that something about our understanding of space and time is off, which, incidentally, is what quantum mechanics is telling us also: that time and space are not what we think they are.

If you were to ask prominent scientists several hundred years ago about reports that people in the countryside were seeing rocks falling from the sky, they would quickly dismiss these as cases of hallucination or otherwise mistaken identity. Everyone, especially the majority of scientists, knew that rocks don't exist in the sky, so how could they fall from the sky?

If you were to ask a majority of physicists even fifty years ago whether they thought the universe was branching every nanosecond into parallel universes with each decision that was made, the idea would have sounded preposterous.

The problem in these cases was very much with the cosmological model that was embraced by most scientists at the time. Expanding (or changing) the cosmological models has led to a better understanding of the universe and our place in it. But changing a prevailing model that is ingrained into the minds of any large group of people who share the same doctrines (religious or scientific) is very difficult indeed.

Fifty or a hundred years from now, might we come to realize that dismissing the Mandela effect out of hand was really because we had the wrong cosmological model all along?

THE MANDELA EFFECT MOVIE—QUANTUM SIMULATION AND TIMELINES

Popular excitement about the Mandela effect seemed to reach a zenith in 2019, the same year that I published *The Simulation Hypothesis* and the 20th anniversary of the release of *The Matrix*. The first feature film about the effect was called *The Mandela Effect*. Although the film is ostensibly about the strange multiple timeline phenomenon, the creator of the film seemed to end up in the same place that we do: that if the Mandela effect exists, it must be because we live in a digital multiverse, like a video game.

In the film, the main character is Brendan, a video game designer, who has lost his daughter, Sam. Both Brendan and his wife are devastated by their loss, though it is Brendan who becomes obsessed. He looks at an old copy of the Berenstain Bears, which he had read to his daughter, and is shocked to see that the spelling had changed. He swears that he remembered it being spelled Bernstein, and he confirms with his brother-in-law, Matt, that they had a conversation about the bears being Jewish.

To his wife's chagrin, Brendan goes down the rabbit hole of the Mandela effect, discovering online many of the effects we have outlined in this chapter. Eventually, he tracks down a professor who confirms that it's possible the effect was happening because of alternate timelines. As Brendan spirals downward, he starts to have alternate memories of Sam being alive, and he becomes convinced that there is an alternate timeline in which his daughter didn't die.

Because Brendan is a video game programmer, he realizes that the Mandela effect must be occurring because the universe as we know it is running as a simulation on a computer—to be more specific, a quantum computer. And as any programmer knows, it is possible to crash any computer program by making alterations. To make a long story short, Brendan finds a quantum computer at his local university and is able to write a program that causes a crash of the simulation and a reset, resulting in a slightly different timeline, the one in which his daughter is still alive.

In addition to visually showing the journey of a newbie getting obsessed by the Mandela effect, this movie brings up some of the same themes that we are exploring in this book. If there are multiple alternate timelines, it makes more sense for this to be a simulated multiverse than a single physical universe, driven by some kind of quantum computing system. If that is the case, then it is all based on information and code, which can be stored, processed, and variables changed and rerun from any point.

Chapter 3

The Simulation Hypothesis—Do We Live Inside a Video Game?

The odds that we're in base reality is one in billions.
—**Elon Musk**, *Code Conference (2016)*

Before we get into details about the larger premise of this book (that we may be living in a simulated multiverse), I would like to lay out some of the reasons why I think we may be living in a simulation, similar to a video game. Both of these premises require a paradigm shift in our understanding of the so-called physical world. This chapter includes a summary of some of the information from my previous book; those who have read it recently can feel free to skim over some of this material.

From Plato and the Vedas to *The Matrix*: The History of an Illusive Idea

The idea that the world around us isn't the *real world* is not a new one. It goes back to the foundational texts of Hinduism, the Vedas, which were composed in India some 5,000 years ago by shadowy figures known as *rishis*. The idea that the physical world is an illusion (or *maya* in Sanskrit) is central to Hinduism and several of its offshoots, including Buddhism.

In fact, this idea that the world around us isn't the real world permeates all of the world's major religions in some form or another, an idea I explored heavily in my previous book (we'll revisit the religious angle in Chapter 14).

Not to be outdone, philosophers both ancient and modern have gotten into the act. After all, sitting around and wondering about the nature of reality isn't limited to either mystics exploring ecstatic states in the Himalayas or college students in late-night dorm discussions. Philosophers seem to occupy an interesting middle ground between scientists and mystics, asking some of the same big questions but combining both logic and insight into arguments that are often hard to refute.

One of the earliest philosophers to write about this idea in the Western world was Plato, whose "Allegory of the Cave" appears in *The Republic*. In it, Plato (through his mentor and literary alter ego, Socrates) explains that most of us are like prisoners inside a cave, chained to a wall in such a way that we can only see the wall opposite the opening of the cave. Outside the cave is a fire or light, which projects shadows onto the wall, like an ancient version of puppet theater. Since the people in the cave have only ever seen the shadows on the wall, that is all they know, believing that the interplay of shadows is reality.

Plato argues that philosophers are metaphorically able to break the chains and escape the cave to see outside. Anyone who does this will at first naturally be blinded by the light (remember, they have spent their entire life in a dark cave), but their eyes will eventually adjust. When they do, the unchained philosopher will see what seem like wondrous things. The philosopher will want to return and tell the people inside the cave what reality is *really* like. But, of course, the people in the cave, not knowing any better, will balk at this idea, not willing to give up their familiar life of being chained inside the cave.

If this sounds a little like the plot of *The Matrix*, it kind of is. Plato's philosopher is giving the two-thousand-year-old equivalent of the red pill or the blue pill choice that Morpheus gives to Neo (see the sidebar if you aren't familiar with the scene or the film). Plato seems to assert that given this choice, most would take the blue pill and go on with their lives inside the dark cave.

Other Western philosophers over the centuries since Plato have weighed in with their own versions of the central idea. René Descartes, the French philosopher and mathematician who created the Cartesian coordinate system, gave slightly different versions of an argument in *Discourse on the Method* and *Meditations on First Philosophy* from 1637 and 1641, respectively. Descartes starts by considering whether there was an evil demon of "utmost power and cunning" that "has employed all his energies in order to deceive me."

If the evil demon could make it seem like Descartes was in a physical world by deceiving his senses, then Descartes could not be sure he was in the real world and not in a dream world. "I shall think that the sky, the air, the earth, colours, shapes, sounds and all external things are merely the delusions of dreams which he has devised to ensnare my judgement."

The only thing he could be sure of, though, was that he was there thinking and perceiving, and this led to his famous phrase, "I think, therefore I am." In another version, there is no evil demon, but Descartes is simply in a dream world that seems, as many dreams do, completely real.

More Modern Versions

This philosophical train of thought was picked up by nineteenth-century philosopher Bishop George Berkeley (after whom the town is named that houses the famous University of California branch across the Bay from San Francisco). Berkeley defended two corners of philosophical thought that relate directly to the simulation hypothesis: *idealism* ("the claim that everything exists either is a mind or depends on a mind for its existence") and *immaterialism* ("the claim that matter does not exist").[39]

These two ideas can be contrasted with *realism* and *materialism,* which both argue that the world out there exists and is independent of the mind. This is, of course, the view of classical science that has become muddied by relativity and quantum physics, which we'll delve into over the next few chapters.

There are also a number of contemporary versions of Descartes' argument, including the Brain in a Vat (B-I-V) scenario and Boltzmann brain scenario.

The B-I-V scenario maps directly to the Descartes argument minus the evil demon. In this scenario, there is an actual brain in a physical vat, with wires that are sending in all the signals that normally come to the brain from the body. The brain, theoretically, should be unable to distinguish between whether it is in a body or not, since it is getting the same signals. The B-I-V scenario is a little more esoteric than even the BCIs (brain-computer interface) used in *The Matrix*, where the entire human being was in a vat with wires plugged into their mind, but along the same idea. Since no one has been able to keep a brain alive outside a human body, this is a thought experiment only, but an interesting one in the context of our discussion about simulations.

The Boltzmann brain argument is a little bit different. It's more of a contemporary physics version of the idea that relies on the magic of infinity, which we'll talk more about in Part II. This argument specifies that in an infinite universe, through quantum fluctuations, all the atoms that make up a normal human brain, complete with memories, are bound to come together at least once, randomly. If they do, then this human brain would think it had a past and was living in a real universe, but it would really have false memories.[40]

Of course, neither the philosophers of hundreds of years ago nor the mystics of thousands of years ago used the terminology of computer simulations or video games. They did, however, use the metaphors of the technology at hand. The Vedas, perhaps the earliest religious texts in the world, used the analogy of a lila, or play, language that was echoed (though in a different definition of the word *play*), whereas Buddhist teachers often have used the metaphor of a dream to describe the physical world.[41]

Video games were in their infancy when Philip K. Dick gave his famous speech in Metz, and it wasn't until 1999 that the idea really caught the popular imagination when several films, including *The Matrix* and *The Thirteenth Floor* and *eXistenZ*, which all tackled the problem of living inside a video game-like world.

THE MATRIX AND THE QUINTESSENTIAL SIMULATION

For those few who haven't seen The Matrix, this is a (very brief) summary. Neo (Keanu Reeves) lives a droll life and works in a droll cubicle in a droll office building. At night, he lives another life online, hacking the net, where he comes across a number of enigmatic references to something called the Matrix. This leads him to Morpheus, named after the Greek god of dreams (Laurence Fishburne), whom Neo asks what the Matrix is.

In what has probably become the most famous scene from that landmark film, Morpheus informs Neo that he cannot tell him what the Matrix is; he must see it for himself. He then proceeds to give Neo the choice of the red pill or the blue pill. If he takes the blue pill, he will wake up in his bed the next morning and nothing will be different. If he takes the red pill, he will wake up. Of course, he takes the red pill and finds that he is a human being that has lived his whole life in a pod, with a wire (what we would now call a brain-computer interface, or BCI) connected to his brain. The wire is used to beam an ultra-realistic multiplayer video game, which is the Matrix, to his mind, and captures his mental responses. The BCI can be used to beam other simulations into his mind, little Matrices, if you will. As Morpheus tells Neo: He has been living in a dream world his entire life.

Why? It turned out that a group of super-intelligent robots had enslaved humanity by plugging them into the Matrix to harness the electricity of the human brain.

The film and its sequels, *The Matrix Reloaded* and *Matrix Revolutions*, are equal parts philosophy, science fiction, and action. They follow the adventures of Neo and company as they try to rescue humanity from this fate, and have become cultural touchstones.

Bostrom's Ancestor Simulations and the Simulation Argument

It wasn't until 2003, when Swedish philosopher Nick Bostrom, now professor of philosophy at Oxford, wrote his landmark paper, "Are You Living in a Computer Simulation?" that the idea moved beyond science fiction and the terminology we use today of the simulation hypothesis started to catch on.

Bostrom starts his now-famous paper with:

> Many works of science fiction as well as some forecasts by serious technologists and futurologists predict that enormous amounts of computing power will be available in the future. Let us suppose for a moment that these predictions are correct. One thing that later generations might do with their super-powerful computers is run detailed simulations of their forebears or of people like their forebears.[42]

Bostrom's argument was about a particular kind of simulation, which he calls *ancestor simulations,* and uses it to make his argument. Let's add this to our growing list of definitions.

> **Ancestor simulation**: Introduced by Nick Bostrom, this refers to a particular kind of simulation run by an advanced technological society that decides to run simulations of their ancestors.

For example, in our own world, this would be as if we were

creating a fully immersive sim that took place in ancient Rome or, if it happened a thousand years from now, our descendants might be simulating us—that is, their 21st-century ancestors.

Bostrom goes on to say that computational power may not be limited on cosmic time scales and that it is possible for technologically advanced civilizations to devote an entire planet's worth of resources toward computing, estimating that would provide more than enough computing power to simulate a whole universe and the people in it.

Bostrom presents his conclusion early on in the abstract, and you can read the paper to get the details:[43]

> It could be the case that the vast majority of minds like ours do not belong to the original race but rather to people simulated by the advanced descendants of an original race. It is then possible to argue that, if this were the case, we would be rational to think that we are likely among the simulated minds rather than among the original biological ones.

The rest of the paper goes into details of how he arrived at this conclusion, which requires at least one civilization to arrive at what he calls post-human, which is analogous to what I like to call the Simulation Point: when a technological civilization can build simulations that are so sophisticated they cannot be distinguished from physical reality.

The Trilemma and the Simulation Hypothesis

Bostrom starts by making a big assumption, that consciousness is really a matter of computation, so that if you have enough computational power, and you can store and process all the memories of a person, you should be able to create a conscious entity, using that information. He calls this *substrate independence*, in that consciousness, which we currently experience via our biological brains, could be simulated in other media, such as silicon. This assumption is, of course, a matter of some vigorous debate.

Bostrom refers to civilizations outside of the simulation as being in *base reality*. He says that civilizations that haven't reached the post-human phase are pre-posthuman, a rather awkward phrase, which is why I have stuck with the Simulation Point.

The argument relies on thinking on cosmic timelines and how advanced civilizations might develop. To consider such a civilization, we must project forward from our current technology to see how we might get to the Simulation Point, which I did in my previous book. This could take a decade, a century, or a millennium, but all of these are blinks of an eye in cosmic time.

In this paper, Bostrom came up with his trilemma, which taken together with the probabilities of each, he refers to as the *simulation argument*. This argument concludes that one of these three scenarios must be true:

1. No civilization ever reaches the Simulation Point.
2. Civilizations reach the Simulation Point but decide against creating ancestor simulations.
3. Civilizations reach the Simulation Point and create many ancestor simulations

According to Bostrom, the probability of these three scenarios should add up to 100%. The conclusions under each scenario are:

1. We are not in a simulation.
2. We are most certainly not in a simulation.
3. We are most certainly in a simulation

What Are the Probabilities That We Are in a Simulation?

If scenario #1 is true, then no civilization ever creates ancestor simulations and the probability that we are in a simulation is most definitely zero.

If scenario #2 is true, then we are most certainly not in a simulation with a high probability, though just because a civilization outlaws creating ancestor simulations doesn't mean that no one will ever create them, since people do things that are outlawed all the time.

Finally, if scenario #3 is true, and one or more technologically advanced civilizations get to this point, they are likely to create many simulations (with many simulated minds), and Bostrom concludes that we are most certainly in a simulation, with a probability close to 100%. The way that Bostrom reached conclusion #3 was through a statistical argument, complete with math.

Bostrom said to take the number of civilizations that have reached this point and the number of simulations they may have created. As long as there is at least one civilization that gets there, they would have created many, many ancestor simulations, since creating one is just a matter of firing up a server and more computing power. If each simulation has, let's say, billions (or trillions) of simulated minds, eventually the number of simulated minds will outnumber the biological minds by a significant margin.

Using simple probability, you can calculate the odds that you, as a mind, are a simulated mind in a computer simulation, as opposed to being a biological mind in base reality. It is basically the ratio of simulated minds to biological minds:

$$\text{Probability That You Are Simulated} \quad \text{equals} \quad \frac{\text{\# of simulated minds}}{\text{\# of biological minds}}$$

Let's suppose there are 1,000 times as many simulated minds as biological minds. The probability becomes: 1000/1 = 99.9%. If there are 10,000 times as many, it become 99.99%, and so on, approaching 100% (but never quite getting there) as the number of simulated minds goes up.

A more simplified version of the argument is often presented on news sites and in public forums. Rather than using probabilities of simulated minds versus biological minds, you could use the ratio of base-reality worlds to simulated worlds. There is only one base reality, but assuming that someone in that base reality reaches the Simulation Point, a civilization could run 1,000, 10,000 or a million simulated worlds. The probability then that we are in base reality would be very small (1 in 1000, or 1 in 10,000 or 1 in a million, etc.).

Bostrom, when asked, generally says that it's difficult to assign probabilities to these three possibilities, but they must add up to 100%, since one of them is most likely true and they are mutually exclusive. In the paper, he says that absent further evidence, we might assume even probability of each of the three possibilities. In one interview, he did give his personal opinion that the likelihood of #3 being true was around 20%.

Video Games and the Simulation

As I mentioned earlier, although the simulation argument had been out for over a decade in the form of Bostrom's paper before I came across it, in my own case it was my background in video games and my speculations of how long it would take for us to reach the theoretical Simulation Point that got me there.

In fact, during the same year that I was playing the VR ping pong game that got my mind rolling, Elon Musk famously said at a conference that the chances that we are in base reality (i.e., that we are not living in a simulation) was essentially one in billions.

Musk was using the video game version of the argument. His point was that if we looked at video games 40 years ago, there was Pong, which was basically a dot and two squares that moved up and down. Today, Musk argued, we have three-dimensional

MMORPGs (massively multiplayer online role-playing games), we have virtual reality and augmented reality, and if we assume any rate of improvement at all, that we will get there, even if it takes 1,000 years or 10,000 years. Then he used the simplified argument about the number of base realities versus simulated worlds to arrive at his conclusion of one in billions.

The core of Bostrom's argument seems to be that if any civilization in base reality ever got to the Simulation Point, we are most likely in a simulation.

Which brings us back to the question that Musk raised and that my experience in virtual reality in 2016 made me ponder: How long will it take the one technological civilization we have as an example (i.e., our civilization) to get to the Simulation Point?

The Road to the Simulation Point

I spent a good amount of space in my previous book detailing the stages of technology development that would take us to the Simulation Point, starting with simple video games and extending into the future. I will summarize them briefly here.[44]

Stages 0–3: From Text Adventures to MMORPGs

The idea of an explorable world inside a computer started with text-based games like *Colossal Cave Adventure* in the 1970s, and reached its peak with the Infocom games like *Zork* and *The Hitchhiker's Guide to the Galaxy.* These text adventures pioneered the idea of a virtual world that could be explored by the player. They also had a gamestate representing the current state of the world in bits; this gamestate was constantly modified as the player's character ventured throughout this world. We'll explore how gamestates are constructed in detail in Chapter 8.

The first graphical game that was widely available, *Pong*, led directly to the arcade and home video console craze of the 1980s, with games like *Space Invaders* and *Pac-Man*. It wasn't until the tools of graphical arcade games were combined with elements of text adventures that we really started down the road to the Simulation Point. These primitive graphical RPGs included *King's Quest*, *Legend of Zelda*, and many others. Although these were simple, 2D, single-player games, they had many of the elements of today's 3D MMORPGs like *World of Warcraft* and *Fortnite* in vastly simplified form: worlds that are rendered and can be explored via characters/avatars that the user role-plays.

In the 1990s, *Doom* broke new ground with a 3D environment that you could explore, and the scene would shift *in real time*. Doom's chief programmer, John Carmack, would later go on to be the CTO at Oculus and contribute heavily to the modern virtual reality boom. Today's MMORPGs house millions of players, using 3D rendering techniques to render only that which can be seen by the player's avatar inside the game, individually on each computer or on the mobile phone. These techniques paved the way for what is being referred to as the metaverse today using more immersive techniques in the next phases.

Stages 4–5: Immersive VR, AR, MR

Building on top of 3D MMORPGs, today's virtual and augmented reality systems are starting to bring science fiction closer to reality.

The modern virtual reality (VR) landscape started to evolve rapidly after the Oculus Rift was introduced and acquired by Facebook in 2012. Many of the other big companies have scrambled to come up with their own virtual reality headsets. In 2018's blockbuster Steven Spielberg movie, *Ready Player One*, for example, characters could not only experience VR through a headset but also use haptic gloves, full-body suits, and even omni-directional treadmills to increase the sense of immersion. Here in the real world, these items are already being developed and, in many cases, are already available on the market as VR headsets become smaller and smaller.

Augmented reality (AR) is less intrusive than VR, because it allows us to see both the virtual and the physical world around us by placing digital objects in the physical world. Today, this is usually accomplished through AR glasses that augment what you see by rendering virtual objects into the world around you. Although initial AR headsets like the Microsoft HoloLens were big and clunky, the size of AR glasses is quickly approaching the size of regular glasses, which means they could be a much more socially accepted way to log in to a virtual world. Less immersive AR techniques have also been built into smartphones, where digital objects are rendered over the real world that is captured by their cameras (a highly successful example was the popular game Pokémon Go).

With AR and VR (which, taken together, are now called XR or MR, for mixed reality), we are getting closer to photorealistic rendering. The number of pixels has increased dramatically from the old 8-bit, 256 colors that I programmed in my first Tic Tac Toe game to the 32- and 64-bit devices of today, not to mention the 4K and 8K pixel screens. The rise of AR shows that the physical world can indeed be modeled, thus starting to blur the line between macro-level objects and information.

Stage 6: Building *Star Trek's Replicators and Holodeck*

Stage 6 includes 3D printers and light-field technology, which represent significant leaps forward in making virtual objects appear (or actually become) physical objects. In fact, these technologies are starting to look more like *Star Trek*'s replicators or its Holodeck than ever before.

First up is light-field technology, which would take AR to a whole new level, not requiring glasses at all but able to make an object appear in the room out of thin air. Light fields are the study of how light bounces off objects and makes its way to different parts of the room or environment. This research can be used to trick the eye (initially with but eventually without glasses) to think there is a solid object in the room when it is essentially a type of hologram. As better and smaller projectors are created, we will be able to use knowledge of light fields to create virtual objects anywhere. As long as you don't try to touch the object, it will look as real as the other objects.[45]

The basic idea of 3D printers is that almost any physical object can be modeled as information and then printed as a series of 3D pixels. Although today's 3D printers can generally only print using one type of ink (usually a single colored thermoplastic), 3D printers have been able to print more complex objects, such as a one-third-scale models of an Aston Martin car, an actual gun, and a number of efforts using a patient's own cells to 3D-print skin that can be grafted on much more successfully over an injury. One startup team actually created a one-third-scale model of a heart, and another is using 3D printing to print blood vessels. The largest 3D printers yet built are being used to assemble objects in orbit. All of this shows that the line between virtual and physical objects is continuing to blur.

Pretty soon, 3D printers will be able to use molecules and atoms of different substances as the pixels with which to print objects, and the line between a physical object and information will be blurred even further. If this happens, like Captain Picard, you'll be able to say, "Tea. Earl Gray. Hot," and have both the cup and the tea fabricated right before your eyes.

Stages 7–8: Mind Interfaces and Implanted Memories

Now let's move beyond where we are today into more speculative areas of technology. One of the main reasons *The Matrix* was so convincing to humans like Neo was that the images were beamed directly into their brains, in this case via a wire that attached to the cerebral cortex. Neo's responses were recorded and sent into the computer system that was running the simulation. As a result, Neo and every other human plugged into the simulation was tricked into thinking the experience was real, like a multi-player dream that never quite ends.

To truly build something like this, we will need to bypass today's VR and AR goggles and interface directly with the brain to visualize the game world and read our intentions of what we wanted our characters to do in the game. Advances made in the past decade suggest that mind interfaces are not as far off as we might think. Startups in this field include Neurable, which is working on BCI for controlling objects within virtual reality, using nothing but your mind. Another startup, Neuralink (funded by Elon Musk), claims to develop high-bandwidth and safe brain-machine interfaces that involve implants, based on a concept from science fiction writer Iain Banks.[46] Facebook and other giants are purchasing BCI companies, though most of them are involved with reading intentions from brain signals (electroencephalogram [EEG]) or muscular response to electrical signals (electromyography [EMG]). So, we are well on the road to being able to read intentions and interpret them. But what about the opposite: broadcasting into the mind?

Experiments done in the 1950s by Wilder Penfield suggest that memories can be triggered inside the brain by electrical signals. But, in what sounds like a science fiction scene out of *Blade Runner*, there are much newer experiments that suggest that memories can also be implanted. In 2013, a team of researchers at MIT, while researching Alzheimer's, found that they could implant false memories in the brains of mice, and these memories ended up having the same neural structure as real memories. This was done in a very limited way, but the techniques are promising.

If memories can be falsified, then we may be entering the world that Stephen Hawking warned us about. Speaking at Harvard, he said: "The history books and our memories could just be illusions. It is the past that tells us who we are. Without it, we lose our identity."[47] Although he wasn't talking about false memories of the type that Philip K. Dick was talking about in his science fiction, we will see that the model we are discussing in this book introduces the possibility that the past is not as fixed as we like to think. [48]

Although we are a ways off from mastering this stage, there is steady progress on BCI technology with different applications, including video game controls, therapeutic applications to allow those who are injured to use electrical signals from the brain to move prosthetic limbs.

Stages 9–10: AI, Simulated, and Downloadable Consciousness

Discussions about artificial intelligence (AI) is pretty common today in Silicon Valley and academia, though today's AI is still pretty primitive compared to that depicted in science fiction books and movies. Take NPCs (nonplayer characters) from video games. Most NPCs today cannot pass the infamous Turing Test. Created by computer pioneer Alan Turing, the test is basically a game wherein a conversation with an AI is indistinguishable from a conversation with a human being.[49]

Even though we don't fully understand consciousness, AI is one of the most rapidly advancing fields in computer science today. Already, AI is giving humans serious competition in traditional games like chess and Go. Unlike Deep Blue, which was the first computer to beat Gary Kasparov in chess, the AlphaGo program didn't rely on rules but learned through self-play. We'll talk about self-play in Chapter 13.

Virtual characters are starting to look more realistic, too. China's Xinhua news agency recently introduced virtual news anchors that can read the news like real humans, and virtual influencers have millions of followers on YouTube. Games giant Epic has introduced Metahuman Creator into its Unreal Engine to make it easy to generate realistic looking faces, and other game companies are following suit. Already, we are seeing AI combined with these virtual characters so that you can talk to them online and feel like you are having one-on-one conversations with virtual characters. [50]

Increasingly, there are those in Silicon Valley, such as Google futurist Ray Kurzweil, who believes that we will soon be able to download our entire consciousness onto a computer. This would be done by mapping what is called the *connectome*, and it has become the subject of a number of science fiction stories, ranging from episodes of *Black Mirror* to *Upload* to Neal Stephenson's *Fall*. Downloadable consciousness is a very big topic, one that was explored in my previous book, and suggests the idea of a digital afterlife. It also gets to the very heart of the role-playing game (RPG) versus NPC discussion and what is now called the hard problem of consciousness.

We'll revisit these ideas of mapping consciousness, loading it (or changing variables in it), and learning from self-play in the next few parts of this book. For now, if we are ever able to download a copy of consciousness fully, we will be able to reach and surpass the Simulation Point.

Reaching the Simulation Point

Just like when I was playing virtual reality ping-pong, where I forgot about the real world for a moment, reaching this point means (1) our virtual environments will become indistinguishable from physical reality and (2) we will be able to simulate virtual characters fully that think they are in a real environment.

And what would it mean if we, as a technological society, are able to get to this point relatively quickly? Would people escape to the virtual world: Why travel when you can have the experience of being in Florence without the jetlag, or fly to India when you can experience the Taj Mahal without the indigestion? Why spend time on real relationships when you could experience almost any type of sex that feels just like the real thing with just about anyone?

As we make progress on the road to the Simulation Point, these are questions that we will have to ask ourselves as a society.

Which leads to a bigger question: Have we asked ourselves all these questions already? Or to paraphrase a popular phrase from the science fiction show Battlestar Galactica, has all of this happened before?

Remember back to Bostrom's startling simulation argument: if any civilization ever gets to the Simulation Point, then we are already likely in a simulation.

NPC versus RPG Simulations

If we are able to progress on the stages on the road to the Simulation Point, this would mean we have gotten to the point of creating what I like to call two "versions" of the Simulation Hypothesis: the RPG version and the NPC version.

This is one aspect of the simulation argument that is often glossed over in popular discussions when referencing *The Matrix* or similar immersive simulation. For Bostrom's statistical argument to work, all, or substantially all, of the beings in simulations would have to be simulated beings and not biological beings playing a simulated being.

Unlike, say, in *The Matrix*, where Neo and Morpheus and Trinity existed outside the simulation, the conscious beings in ancestor simulations would be just programs and data. In the world of video games, we refer to these as NPC's, like the orcs in a fantasy-themed game that you might fight. NPCs are common in video games, although they are not very sophisticated yet; we have not really married some of the advancements of AI to virtual characters other than to their appearance, which has gotten more and more realistic.

When we play a video game, we exist outside of the game, and we are role-playing a character, or an avatar, inside the game. This is, of course, closer to what is depicted in *The Matrix,* an RPG simulation.

You will notice that these aren't totally mutually exclusive; you could be an avatar inside *World of Warcraft,* for example, role-playing your avatar, and you might also see many creatures like orcs, who are NPCs being controlled by the game engine. Agent Smith in the Matrix was an AI within the RPG simulation.

Although both versions can coexist, Bostrom's math relies on there being way more NPCs (unless the biological beings can role-play thousands of characters).

Nevertheless, Bostrom's argument, though convincing and responsible for bringing simulation theory into a respectable discussion, isn't the only one that implies we might be living inside a computer-generated world. Philip K. Dick came to the same conclusion on different grounds, one of multiple timelines, and most of the philosophical constructs I mentioned at the start of this chapter—Plato's cave, Descartes' evil demon, Berkeley's idealism, and even the Brain-in-a-Vat—would all be closer to the RPG version of a simulation: that you (or the brain) is real, but everything that you are perceiving is not.

The RPG vs. NPC debate parallels a very old, on-going dialog about consciousness. Is consciousness a result of the material world (i.e., the brain and neurons) or does consciousness exist outside the material world and only inhabits our bodies? It lies at the heart not only of the religion versus science debate but within science itself, particularly in physics (some of which we will get into later in this book), which actually birthed the multiverse idea as an interpretation of quantum mechanics to avoid having consciousness or observation be considered a key part of the material world.

For this reason, whereas Bostrom uses the term *simulation hypothesis* to apply only to conclusion #3 in his simulation argument, the term has taken on the broader meaning of representing the idea that we may be living, as Philip K. Dick proposed and *The Matrix* visualized, in a computer-generated, simulated reality.

This RPG version of reality inevitably sparks comparisons to religious beliefs (which I covered in depth in my book, *The Simulation Hypothesis*) for Eastern religions like Buddhism and Hinduism as well as for the Western religions, Christianity, Judaism, and Islam. Rather than getting into these theological implications here, I will discuss what this all means in the context of a simulated multiverse in Chapter 14.

Where We Go Next

From here, we will turn our attention from the science fiction concepts of Part I to the aspects of hard science that nevertheless sound like science fiction. What does simulation theory have to do with physics and the multiverse? Some of the baffling findings of modern physics, including the mysterious observer effect, the many-worlds interpretation, and the delayed-choice experiment, all suggest that there is more to space and time than meets the eye.

Although I covered some of this material in my previous book, in this book I want to focus on how these findings suggest not just that we are in a simulated universe, but that we are in a simulated multiverse. We'll explore the latest scientific thinking on multiverse, quantum mechanics, and time in the next part of this book.

THE THIRTEENTH FLOOR AND NESTED SIMULATIONS

Released in the same year as *The Matrix*, *The Thirteenth Floor* was based loosely on a German film, *World on a Wire* (1974), which was based on the novel *Simulacron-3* from 1964.

The film starts out in 1937 Los Angeles with a gentleman, Hannon Fuller, leaving a note with a bartender at the Wilshire Grand Hotel, intended for his colleague Douglas Hall, about something incredible he'd discovered. We quickly realize that the 1930s LA is a simulation and the world outside is the contemporary world of the movie's release date, the Los Angeles of 1999. As Fuller comes out of the simulation, true to its noir thriller feel, he is murdered.

Fuller is actually the owner of a VR company that has built the realistic simulation, which contains virtual simulacra, or what we might call NPCs today. These NPCs live their entire lives without knowing they are in the simulation. Fuller's protégé, Douglas Hall, is the focal point of the movie, and he is questioned by LAPD detectives after Fuller's death. Somehow, Fuller's daughter, Jane, shows up to take over the company and shut down the simulation.

Hall is being blamed for the murder of Fuller and decides to enter the simulation when he realizes that Fuller left a message for him. We now see the extent of the simulated VR world that is indistinguishable from physical reality as Hall plays the role of a bank clerk, John Fitzgerald, who was living his life in 1930s LA. Hall takes over this avatar as his user, while Hall is jacked into the simulation. After Hall logs out, John Fitzgerald becomes an NPC again living his life in the simulation.

Through some intrigue, a bartender who was supposed to deliver Fuller's message to Hall, read it instead. The message, in essence, said that if you drove to the edge of the city, you could see that you were in a fake world. After seeing the edges of the sim outside the city, Hall exits back to 1999.

At this point (spoiler alert), things get weird. The LAPD detectives inform Hall that there is no Jane Fuller and there never was—that Fuller never had a daughter to begin with. This is confusing, and Hall tries to figure out what's going on, at one point finding a woman who looks just like Jane, but is not her. She is rather, a grocery clerk and has no memory of meeting him.

Hall then proceeds to take Fuller's advice from his note: to drive out somewhere that he would normally never go. He does, only to find the edges of a VR simulation: It turns out that 1999 LA was also a simulation. Jane was actually a user from outside the simulation (i.e., from a future in the real world) who inhabited the grocery clerk's body, just as Hall had done with the bank teller from 1937. Jane eventually informs Hall that thousands of simulations are running, but that his was the first one that had developed a simulation within it (which are often referred to as stacked or nested simulations).

In a further set of twists of consciousness and body hopping across the 1937 and 1999 simulations, Hall's consciousness eventually escapes the simulation and wakes up in the future, in 2024, and meets the real Hannon Fuller and his daughter, Jane, on whom the grocery clerk's avatar was based.

Not only is this vision of a simulation close to what we might think of when we create virtual, simulated realities, complete with many simulations and large numbers of NPCs and a smaller number of RPG players, it also touches on the stacked-simulation scenario and why it might be difficult to allow too many levels of simulation, because they would be drawing on the same computing power. The creators of the simulation might need to shut down the nested simulations

Part II

Some Far Out Science

Nobody understands quantum mechanics.
                    —Richard Feynman

Time does not exist—we invented it.
                    —Albert Einstein

Chapter 4

A Variety of Multiverses

I believe we live in a multiverse of universes.

—Michio Kaku

In this part of the book, our goal is to explore the science related to quantum mechanics and the quantum multiverse. First, though, in this chapter, we'll survey the different types of multiverses. Then, we'll have a chapter devoted to one of the core mysteries of quantum mechanics, often called quantum indeterminacy or the observer effect, and relate it to the simulation hypothesis.

We'll then move on to the many-worlds interpretation (or MWI) of quantum mechanics that provides the scientific underpinnings for the multiverse thesis that I'm exploring in this book. Finally, we'll end this part of the book with a chapter on how quantum indeterminacy is even stranger than it looks, because it shows us that time (the past and the future) are not what we think they are. Eventually, we'll use this knowledge to explore how a simulated universe and multiverse might work, replete with multiple parallel timelines and multiple possible futures and pasts.

The Multiverses, According to Sci-Fi

As I said in *Chapter 2: The Mandela Effect—Real or Mass Delusion?*, science fiction involving parallel universes seems to be growing significantly in the twenty-first century. This parallels the popularity of space travel and time travel stories in the past century.

Let's start by looking at the way science fiction writers have depicted parallel universes over the past few decades. Many of these examples are explored in more detail in the sidebars throughout the book, so this list will give just a short overview of each:

Sliding Doors and **Run Lola Run** (1998)—These two movies, neither of which is technically science fiction, were released in the same year. We see the idea of timelines branching from a single point which lead to different outcomes. In the example of *Sliding Doors*, a separate timeline branches off of the first timeline and then exists in parallel for some time, overlapping the main timeline, before merging back in. In *Run Lola Run*, on the other hand, we see Lola trying to rescue her boyfriend Manni by rewinding what happened and making different choices multiple times. We see visually what running our Core Loop might look like in a real-world, high-stress situation.

The Thirteenth Floor (1999)— As mentioned in the previous chapter this film is one of the best representations of ancestor simulations. When the protagonist finds out he is living in a simulation, one of the RPG players, who exists outside the simulation, tells him that their simulation is "one of thousands." The thing that makes this simulation unique is that it is the only one where they in turn develop their own ancestor simulations, or nested simulations. Although we see only the nested simulations and not the parallel ones, they are definitely there, which means that it is also faithfully representing a simulated multiverse.

Fringe (2008–2013) and **Counterpart** (2017–2018)— In the twenty-first century, two popular TV shows demonstrate the idea of a single parallel world that has somehow split off from this world, but retains many similarities, including a shared history. The source of the divergence is never explained fully, but the existence of a parallel world with alternate versions of the main characters is a key plot point in both. Both shows reveal that some physics phenomenon was responsible for either (1) breaching a way into the other universe or (2) causing a branch off the main universe to create the second one.

Sliders (1995) —The science fiction series *Sliders* is a more direct example of navigating a quantum multiverse. In *Sliders*, wormholes are used to visit other universes, each of which deviates in some way from our own, many with different versions of the main characters. How do they develop the wormhole? Quinn, the main character, is shown how by a different, parallel version of himself, a theme we will see repeating in other multiverse stories.

Dark Matter (2017) — In a similar manner, the hero of the novel *Dark Matter*, Jason Dissen, a failed quantum physicist who is happy with this life, encounters an alternate version of himself. This alternate version was more successful as a physicist and developed a machine which can put large objects into superposition (a concept we'll explore heavily in the next few chapters). This device results in an ability to go to different universes and encounter alternate versions of well, everyone. The other Jason, the brilliant one, is keen on stealing the hero Jason's happy home life. Chaos ensues. This is well worth a read if you are inclined to read novels and want to consider the possibilities of multiversal travel.

The Flash/Arrowverse (2010s) and the Marvel Multiverse (2020s)— In *The Flash* and *Arrowverse*, the team gets to other Earths by teaming up with billionaire physicist Harrison Wells, who is running a particle accelerator. Not only does an experiment that went awry allow for the development of super abilities, it also allows for time travel and travel to alternate

Earths (from Earth-1, the main universe). The plotlines involving these multiple Earths are somewhat complicated, but we often see colorful versions of the main characters/superheroes and, in some cases, new superheroes that don't exist on Earth-1.

Not to be outdone by its major comic books rival, Marvel has used streaming TV shows to introduce its own idea of a multiverse. In 2021, the limited series *Loki* introduces the Time Variance Authority (TVA), whose mission it is to prune off multiple timelines before they have a chance to blossom into a full multiverse. These timelines are all depicted on an old television type screen as an on-going tree like structure. Variant branches pruned before they become too big. Without the TVA to enforce a single "sacred timeline," Loki is told eventually, we would get chaos and a multiversal war. This is the starting point for the Marvel "multiverse of madness" which will continue in many tv shows and movies, showing once again that the multiverse has successfully passed the ten-year-old test.

There is an important point in most of these stories: the main characters are able to interact with one or more parallel universes. How is this done? Although this isn't usually explained in too much detail, it often involves some layman's use of physics terms (a particle accelerator, a wormhole generator), or in some cases, a super hero or psychic ability to traverse universes.

SLIDING DOOR AND OVERLAPPING TIMELINES

In the 1998 movie, *Sliding Doors*, we find a vivid example of multiple timelines that branch out and then merge again. When branched, it's as if the two branches are occupying the same time, and each protagonist (or rather, each version of the protagonist, played by Gwyneth Paltrow) sees the versions of the other characters and sense that something else might be going on.

The name of the movie is drawn from the premise that sets the alternative timelines into motion: one day in London, Helen (Paltrow) gets fired from her public relations job. On her way home, as she tries to get on the subway (the Tube), we magically see one version of her make it onto the train while the other version gets caught just outside the sliding doors, missing the train.

This sets off two intertwining timelines. In one of these timelines, Helen sits next to James, who strikes up a conversation with her, and then she arrives home early to find her fiancé, Gerry, in bed with another woman. In a very British kind of way, she makes a joke about them both getting "sacked" that day.

In this timeline, she breaks off her relationship with Gerry, and sets out on her own, starting a PR firm and meeting James again, with whom she starts a relationship. The movie distinguishes between the two Helens visually by having the independent Helen change her hairstyle.

Meanwhile, back in the original timeline, Helen suspects that something is wrong with Gerry, but she arrives too late to catch him in the act. After a series of events, she finds that she is miserable with Gerry in this timeline.

The movie does a good job of trying to show what might happen if there were a macro decision point where the timelines branch off. I won't give away the whole plot, but let's just say that at the end, the timelines merge so only one of the timelines is left. Yet, as in many parallel universe stories, there's a little bit of memory that seems to seep through from the non-surviving timeline!

The Different Types of Multiverses, According to Scientists

In his book, *The Hidden Reality*, Columbia professor of physics and mathematics Brian Greene gives us an overview of nine possible types of parallel universes, each of which results from different theories in physics. Greene gives colorful names to them, such as the quilted universe, inflationary multiverse, and the quantum multiverse; he even refers briefly to a simulated multiverse. MIT Professor of physics Max Tegmark covers similar ground in his papers and books, though his classification uses a numbering system, labeling each multiverse with a level from I-IV.[51]

For our discussion, I have boiled down the list of possible multiverses to five types.[52] We only have time to do a quick tour of these other four types of multiverses before diving deep into the quantum world in the next chapter.

Type 1: Black Holes and Wormholes as Gateways to Other Universes

Whereas Einstein revolutionized our understanding of time and space in his special theory of relativity, it was his general theory of relativity, published in 1915, that really made waves in cosmology. Einstein told us to visualize space-time as a flat surface, for example a tablecloth that is flat and uniform when there are no objects on it. When objects are put on it, they cause indentations in the tablecloth (or think of a bowling ball on a foam surface). These indentations produce curvature in the fabric of space and time. Gravity, said Einstein, is a result of this curvature. Theoretically, he predicted, this curvature of space-time should bend light. It was British astronomer Sir Arthur Eddington who confirmed the predicted bending of light in an expedition during a solar eclipse in 1919, instantly making Einstein the most famous scientist in the world.

Further confirmation of the general theory came throughout the 1920s. One aspect of general relativity, which arose from astronomer Karl Schwarzschild solving Einstein's field equations, was that an extremely dense star would have a very high gravity field that made it difficult for anything to escape. This resulted in a prediction of a singularity, a point that has such high density that it causes a rip in the fabric of space-time itself.

Figure 3: First ever image of a black hole (source: NASA)

This area was referred to as a black hole by John Wheeler in the 1960s and this name stuck. Although black holes have become very popular in science fiction, there was a time when most scientists thought they didn't exist but were just a mathematical curiosity resulting from Schwarzschild's solution to Einstein's equations. The equations show a theoretical sphere around the singularity that is part of the black hole, and anything that passes this point, including light, is unable to escape the gravity of the singularity. The boundary of theoretical sphere is defined as the event horizon, which is defined by the Schwarzschild radius from the singularity outward. The first black hole wasn't discovered until 1964 (dubbed Cygnus X-1) using X-ray astronomy. The first black hole to be photographed occurred in 2019 using data gathered from telescopes all over the earth (shown in Figure 3). [53]

So, what do black holes have to do with parallel universes?

Some scientists believe that if you go inside a black hole, you have the possibility of entering a wormhole, which was more formally defined as an Einstein-Rosen bridge. Where would this wormhole take you? The answer is up for debate, but many physicists are concerned that because of the warping of time and space that happens in a black hole, there could be bridges to anywhere and anytime.

Some say it would take you to another part of our universe. A common trope on *Star Trek* is the existence of wormholes that can open up and transport a starship to a faraway part of our galaxy. (This is basically the setup of the plots of both *Star Trek: Deep Space Nine* and *Star Trek: Voyager*, both of which came out in the 1990s.)

In his book, *Parallel Universes*, physicist Fred Alan Wolf tells of meeting Dr. Martin Kruskal at UCLA in the 1960s. Kruskal, a highly respected mathematician who worked on the Manhattan Project, created maps that showed that a black hole might be a doorway to a parallel universe, though a different kind of parallel universe than the ones we have been exploring in this book.

When Kruskal tried to draw maps of space-time inside the Schwarzschild radius of a black hole, he found that because time slows down to a standstill, there were actually two singularities, one in the past and one in the future, with time going in reverse from one another (i.e., forward in a black hole and backward in a white hole). Whereas John Wheeler, whose exploits will be featured in subsequent chapters, coined the term *black hole*, the term for white holes came into general usage by physicists looking to name Kruskal's opposite singularity. Australian mathematician Roy Kerr took Kruskal's solution (which was for a stationary black hole) further by coming up with a solution for a rotating black hole that showed that there was "an infinite patchwork of parallel universes," which could be formed via black holes and reached through wormholes. These are depicted today in what are called Penrose diagrams, as shown in Figure 4.

These solutions were (and are), of course, theoretical—no one has been able to show that a black hole goes to one parallel universe, let alone many such universes. Even if black holes behaved in this way, it wasn't clear whether it would be possible to travel through them—that is, whether they are traversable.

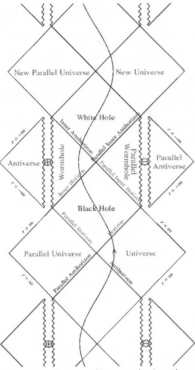

Figure 4: A Penrose diagram depicts the parallel worlds that might be reached via a rotating black hole.[54]

Perhaps more interesting from our perspective is that a black hole's gravity distortion (more accurately, the singularity in the center of the black hole) is of such magnitude that it not only curves space but correspondingly curves time. In the movie *Interstellar*, for example, a ship that goes to a planet near a black hole finds that they are subject to significant time dilation effects.

Some scientists believe that a black hole or singularity could make it possible to create what are called *closed time-like curves* in space-time. This would be a curve that allows you to go back to some point in the past. Although Einstein's equations don't rule these out, physicists have looked into ways that they might be created to travel through time. Kip Thorne and some of his colleagues at Caltech, for example, have worked out a framework whereby wormholes could be used to send an object on such a closed time-like curve.

Wormholes and black holes are interesting, of course, but this type of parallel universe is one that we can only dream about at the moment because of the difficulty of accessing the gravity of a black hole. Moreover, the difficulty in creating a traversable wormhole that would stay open and not collapse immediately as soon as it was formed is another unknown. It is something that, theoretically, an arbitrarily advanced civilization might be able to do, since Einstein's equations don't rule it out.

There are physicists who believe that we will be able to artificially create small black holes or wormholes at some point in the future. This would allow us to tunnel through to a different universe or spawn new timelines, just some people have accused CERN of doing and causing the Mandela effect.

FRINGE AND A SINGLE PARALLEL UNIVERSE

Fringe, which premiered in 2008, was kind of a twenty-first century spiritual successor to *X-Files*. It centers on the Fringe Division, a task force primarily of the FBI and Homeland Security, that is led by agent Olivia Dunham, who turns to a former Harvard professor named Walter Bishop. Walter has spent the past decade in a mental institution, and Olivia recruits his son, Peter Bishop, to help get him out. What starts off as a fringe/creature episode of the week-type of show in which the team is investigating strange cases, turns out to have a larger storyline that involves parallel worlds. It revolves around work done by a mysterious firm called Massive Dynamics, which was started by Bishop's old lab partner, William Bell (an aging Leonard Nimoy, in one of his last roles).

The twist comes in because Dr. Bishop (Walter, the father) has lost memories of what he was working on (possibly with Bell) before he was admitted into the mental institution. Through a series of flashbacks, we find out that doctors Bishop and Bell had together invented a viewing device, like a flat TV screen that could view into an alternate world, where things were similar to how they are in our timeline, but a little different. For example, the Hindenburg never blew up in that timeframe, and the Empire State Building, which was built to have a port for airships at its apex, is still a bustling airship port. As the series moves on, we learn that not only was Bishop able to travel into this alternate world, but that his son, who died from a terminal medical condition, was still alive, though sick there. Walter had finally devised a cure, but it was too late for his son, at least in our timeline.

In an example of the kinds of sins that often play out in parallel interacting worlds stories, given that his wife (in our reality) was upset about the loss of Peter, Walter brings over the other Peter from the other universe and cures him. This of course, causes havoc in the alternate universe, where Peter Bishop is thought of as the Lindbergh baby who was kidnapped and never found. Eventually, each of the team member meets their doppelgangers in the other world: Walter's is the Secretary of Defense, Olivia's has a completely different personality, and Peter, at least at first, strangely is nowhere to be found in the other universe. These alternates are cleverly nicknamed Walternate and Faux-livia.

Fringe has a complicated storyline (involving shapeshifters, experiments on children with psychic powers that allow them to open up the door between the universes, not to mention curry-loving bald aliens called Observers). Perhaps most relevant here is that it serves as a one of many cautionary examples of a bridge between the universes that leads to problems. This includes not only the problem with Peter and Walter, but structural problems in the other universe as well resulting from the breach.

Type 2: Expanding Bubbles (with Doppelgangers)

The next two types of parallel universes have to do with cosmology and the idea of an inflationary universe. They also have to do with what I like to call the "magic of infinity." When something can't be explained, one mental trick that physicists and philosophers sometimes use is to imagine infinity, a number that is so large that it can't actually be imagined. Because it's so large, even if you were to subtract infinity, or double infinity, you would still end up with ... infinity. In that sense it's a special number, like zero, and not a number at all but a placeholder.

This type of multiverse arises from simply assuming that our physical universe is infinite. Now, the farthest distance that we can look is approximately 13.7 billion light years away, which cosmologists put as the age of our universe based on how far the cosmic background radiation has traveled to us. They estimate this to be the time since the Big Bang, currently the most popular theory about the origin of the universe.

In addition, physicists have found that other galaxies are moving away from us and from each other. In fact, they are moving away from us at such a pace that at some point, we will no longer be able to see them because we are limited to our ability to see only those from which the speed of light can reach us. There are parts of the universe that, if the current expansion continues, we *were* able to observe at some point but will no longer be able to see because their light will never reach us.

This is the first mental trick and what I think of as the simplest type of parallel universes: bubbles of observable universes in a much bigger, infinite space that is our physical universe. If the universe is infinite, then if we take infinity minus the size of our observable universe, how much space is left? It's still infinite! This means that there is an *infinite* amount of space that we can *never* observe. In that space, the assumption is that there will be every configuration of particles that can fit inside a bubble that is roughly the size of our observable universe.

Observable Universe

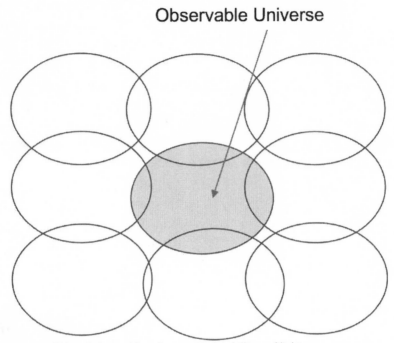

Figure 5: Bubbles of observable universes. Each is billions of light years across

This is why this version is often called the Doppelganger universe[55], because if you think of all the particles in the observable universe, let's say a big number like 10^{80}, then as you go farther out, there would be another observable universe of the same size, like another bubble outside of our bubble. Each bubble might have 10^{80} particles, but the number of total particles is, essentially, infinite.

The idea is that somewhere in these other bubbles of observable universes (even though they are all in the same physical infinite universe), a set of particles will be assembled just like our Earth and there will even be a version of you, perhaps reading a different version of this book. Given a finite number of particles and infinite number of bubbles of that many particles, each pattern will repeat somewhere, kind of like the Boltzmann brain argument we discussed in Chapter 3.

To be honest, I find this version of the multiverse rather unconvincing, or at the very least, unlikely to lead to the kinds of multiple timelines with doppelgangers that we are exploring in this book.

One problem with this type of multiverse is that there would need to be empty space between the universes; otherwise, those at the edge of one universe would be able to see into the adjacent universe, which means they are not so distinct at all and overlap, as shown in Figure 5.

The chance that there really is another me would rely not just on the particles being exactly the way they are in our universe bubble, but also on a very similar history. Even proponents of this type of multiverse agree that they will have a different history from those in our universe, even though it evolved from the same part of the Big Bang, which means that it has the same set of physics rules and constants that we have. Will there be another version of my mother and father and me that has the same history as me?

Unlikely, because even small changes in initial conditions can cause great variance in the results, a phenomenon that is studied in complexity and chaos theory. The counter argument is that *infinity* takes care of that by saying even if that version of Earth doesn't have you but has me, there is always another doppelganger universe somewhere out there that has both of us, through the magic of infinity. There would be an infinity of universes with particles exactly like ours, with an almost infinite number of me and an infinite number of you, each of which has a different history.

Complexity theory shows us that some algorithms produce stable configurations, whereas others produce chaotic ones, which means that they don't repeat in a predictable way. There are many assumptions about this type of parallel observable universe, but it certainly brings us some strange philosophical and scientific questions and feels a bit like hand waving. Infinity becomes the new God, able to wave its magical wand to make every configuration appear an infinite number of times.[56]

Type 3: Inflationary Bubbles

In Type 2, we just need to assume that the universe is infinite and that each bubble universe (which in this case means observable universe) evolved from the same set of initial conditions (quantum fluctuations). It also relies on expansion to a certain extent, because it means that what is not observable now will never become observable. It also means that the laws of physics are similar to our universe's bubble; we just can't observe these other bubbles because they are too far away from the perspective of the speed of light.

There is another version of bubble universes that provides for weirder bubbles with different laws of physics than our own. In fact, the other universes in this model may be nothing like our universe.

This type of multiverse relies heavily on the idea of cosmological inflation, which is defined as a phase of exponential growth that came after the initial blast of the Big Bang. This inflationary period would have had much faster growth than the simple expansion we are seeing in the universe today (where galaxies are moving away from each other). I like to think of the difference between inflation and expansion like the different speeds at which you can inflate a balloon. If you blow air into the balloon manually yourself, it expands slowly (expansion) and eventually gets to the intended shape as long as you keep blowing. However, if you hook it up to a machine, as professional balloonists do, the expansion happens very rapidly, almost instantly, as it transforms from a limp, empty balloon to reveal its fully formed balloon shape.

The idea of cosmological inflation, or just inflation for short, was proposed by Alan Guth, now a physics professor from MIT, while he was working at the Stanford Linear Accelerator with his colleagues as a way to solve a number of problems with the existing theories of the origins of the universe. [57] The basic idea is that this inflationary period began very soon after the Big Bang (in this case, I mean *very* soon, from approximately 10^{-36} seconds after the to 10^{-32} seconds). That means the whole process of cosmic inflation started and ended before a single second had passed from the Big Bang! The insight that Guth and his colleagues had was that there was a period of repulsive gravity.

As mentioned earlier, in Einstein's general theory of relativity, space-time is pictured as a two-dimensional tablecloth (or foam, perhaps), and objects placed there create an indent in the fabric of space-time. This indent defines how other objects that are less massive or more massive would interact with it. During a period of repulsive gravity, rather than indenting, the whole fabric would be spread out very quickly as everything moved away from each other. Once the fast inflation was done, the universe resumed a more gradual expansion, like what we see today.

Guth and his colleagues, Andrew Linde of Stanford and Alexander Vilenkin of Tufts, realized that this inflation wasn't just a one-time event. In fact, every time it happens, this stretches out a tiny bit of space-time into a very large cavern of space-time. This cavern is perhaps as large as our universe! In other words, every time this process happens, a brand-new universe is created, and this universe is not accessible to the residents of the previous universe.

In this version of the multiverse, which Greene calls the inflationary multiverse, each universe is a distinct region resulting from the initial Big Bang. However, because the inflation happens so rapidly, it amplifies whatever fluctuations (called quantum seed fluctuations) were happening in the little bit of space that started the whole process going.

Unlike in the more static, endless expansionary doppelganger universe, where we are relying on distances to separate the universes, each universe is literally inaccessible to the others. These universes are divided not by empty space, which could be traversed, but by a gulf that cannot be traversed, even if you were going faster than the speed of light, because they may rely on different laws of physics.

Greene uses the analogy of a large block of Swiss cheese as the original universe that results from the Big Bang, and each time cosmological inflation happens, it creates a new hole in the Swiss cheese. In this model, each hole is thought of as a bubble or pocket universe.[58]

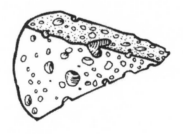

Figure 6: Bubbles in Swiss cheese is a way to visualize inflationary bubble universes. [59]

What is between the holes? Using this analogy, the area in between the holes (i.e., the cheese) is still unstable and may undergo inflation at any time, creating a new bubble universe. In fact, it may not even have the same physics as the other universes do. Each newly created universe is a parallel universe but with different values for many of the constants that we take for granted.

Fine-Tuning and the Multiverse

This version of the multiverse is one of the types that are better equipped to tackle a mystery scientists call the fine-tuning principle (along with the version we will cover in Chapter 6).

Put briefly, the fine-tuning principle is based on the observation that there are a handful of characteristics about our universe that just so happen to be perfect for the formation of galaxies, planets, and of course, for the emergence of our kind of life. Because the conditions are ideal for our kind of life, it's also referred to as the anthropic principle.

In the inflationary multiverse, each bubble universe contains galaxies and planets and everything else we see. Recall that each bubble (or pocket universe) was the result of quantum seed fluctuations when the initial universe was quite small, and each region of rapid expansion created its own bubble.

However, if any of these original variables or seed fluctuations were different, we might end up with a very different kind of universe, with very different kinds of physical constants. Here are some examples of fine tuning in our universe:

- MIT physicist Max Tegmark and British astronomer Martin Rees calculated that if the cosmic seed fluctuation amplitude, which was approximately .002%, varied and became smaller, galaxies would never form.[60] If it was much larger, there would be other difficulties, resulting in a much larger volume of collisions.
- According to Rees, the ratio of the electromagnetic force to the gravitational force between protons is fine-tuned to be approximately 10^{36}. If it was smaller, then only a small, short-lived universe could exist.[61]
- If protons were just .2% heavier, they wouldn't be able to hold onto electrons.[62]

- If the electromagnetic force was 4% weaker, our sun would immediately explode. If the force was stronger, carbon and oxygen atoms would no longer be stable.

- The Hoyle state, which is an energy state of the carbon-12 nucleus, has an energy of 7.656 MeV above ground level. If the state's energy were lower than 7.3 or greater than 7.9, there wouldn't be enough carbon to support life as we know it, which includes human beings.[63]

- Dark energy is what physicists call a force that's causing our universe's expansion rate to accelerate over time.[64] Although we don't know what it is exactly, we can do calculations of it; its density is $\sim 10^{-27}$. Tegmark says that if the density of dark energy could be tuned by a knob from the minimum possible value (10^{-97}) to the maximum possible value (10^{97}) and you were trying to find just the right value, it would take a Herculean effort to get the value just right.

In this version of the multiverse, we see why a multiverse provides a rationale for fine tuning and the anthropic principle. Since each universe might have its own values for things like gravity, cosmological constants, percentage of dark matter, and so on, each universe might have completely different rules of physics. In fact, this is one of the more attractive reasons for the fine tuning.

Max Tegmark points out that there really are only three possible reasons for fine tuning that we can think of:[65]

1. ***Fluke***. The fine tuning of our universe is a fluke and just a coincidence.

2. ***Design***. The fine tuning of our universe is by design; it was created for life. Who could have done the creating? This would be advanced beings who could be literal deities or, simply from our perspective,

deities. Tegmark includes the idea that we might be in a simulation that was fine-tuned for our life.

3. **_Multiverse_**. There are an infinite number of universes, and one of them just happens to be fine tuned for our kind of life, and that's the one we are in.

I (and many others) think that explanation #1, that the whole universe is a fluke, is unlikely (though, of course, we can't prove that it wasn't—at least not yet!). Which leaves #2 and #3, Design or the Multiverse. Some would say that these two are mutually exclusive, that you could have a creator or creators, or you could have an infinite number of universes in a multiverse, but, as the old saying goes, never the twain shall meet.

The whole point of this book is that there is a scenario in which both #2 and #3 are true. We are in a simulation, and it is one of many, many simulations being run on some advanced race's computer system. The only simulations that are allowed to run beyond a certain point are the ones where life arises. (These may or may not be ancestor simulations; in fact, the beings doing the simulations might be completely different from us.)

In a simulated multiverse, we wouldn't need to have an infinite number of dead universes (where no life exists) hanging around. Like evolution, whatever is driving the creation and random changes that occur between universes would also be responsible for pruning the tree of new universes to those that actually meet its criteria. Only those that could sustain, for example, planets and stars and life would be allowed to run beyond a certain point.

Even if we take #3 to be the case on its own, although bubble universes might be able to tackle the fine-tuning principle, we are still left with all of the strangeness of quantum phenomena in our universe and no good explanation for why things work the way they do. Moreover, the bubble universes covered thus far have completely different timelines, so it's unlikely to provide an explanation for the kind of timeline divergence and merging that we are exploring in this book.

Type 4: Universes in Higher Dimensions

A fertile area for alternate parallel universes is that they are hiding in higher dimensions. This idea is based on string theory, which is yet another conjectured theory of how the universe works. Einstein himself couldn't find a unified theory that pulled together gravity along with all the other forces in nature (electromagnetism, gravity, weak nuclear force, strong nuclear force). Quantum field theory, on the other hand, could explain and predict three of these with the exception of gravity, which means that it is also incomplete.

String theory (more formally, superstring theory) came around in the 1980s and was lauded as a potential theory of everything, which could integrate Einstein's relativity and quantum mechanics. Whereas quantum mechanics showed that everything had fundamental quanta, or particles, in string theory, even these particles are really small vibrating strings. The rate of vibration of the strings allows for the differences in particles.

One by-product of string theory is the need for additional dimensions beyond the usual three (or four if you consider space-time). In fact, variations of string theory require 10 or 11 dimensions and a type of multiverse that has been discussed by physicists, one in which there are universes hiding inside these nearby dimensions.

Despite the initial enthusiasm it gathered, today string theory is regarded with some suspicion by some scientists, although others believe it still holds the key to a unified theory. It is hard to come up with experiments that confirm any of its predictions. In the end, it is a lot more speculative and controversial than either general relativity or quantum field theory, both of which have produced experimentally verifiable results.

For that reason, and the fact that the extra dimensions implied by string theory aren't exactly germane to our own discussion of multiple timelines, I won't spend a lot of time on this version of the multiverse. However, the core idea that is interesting to our discussion is that extra dimensions are hiding out, just beyond our purview, and this idea can help us visualize a larger coordinate system in which multiverses might exist.

Hyperspace and Hypercubes: Living in Extra Dimensions

What would these other dimensions and the universes embedded within them look like?

Well for one thing, since Einstein, we are used to talking about time as the fourth dimension, and we'll be using it when describing how multiverses evolve geometrically. But the history of dimensions is an interesting one, and is recounted by Dr. Michio Kaku in his book, *Hyperspace*.

It was the Greeks who first stated emphatically the point of physical dimensions being limited to three. Euclid, the founder of Euclidean geometry, which we all studied in school, stated that a point has no dimension, a line has one dimension (length), a plane has two dimensions (length and width), and a solid has three dimensions (length, width, and depth). Aristotle and, later, Ptolemy from Alexandria agreed with Euclid but went further to state that there were no other dimensions. Aristotle writes in *On Heaven*: "The line has magnitude one way, the plane in two ways, and the solid in three ways, and beyond these there is no other magnitude because the three are all." Ptolemy went further and offered a proof that there are no higher dimensions.[66]

But in the nineteenth century, this idea of a limited set of dimensions started to be questioned. Georg Friedrich Bernhard Riemann, in particular, shattered the idea that mathematics would be limited to only three dimensions and no more. This gave rise to the Riemann geometry and the idea that any shape could be extended into extra dimensions. Riemann pointed out in particular that the Pythagorean theorem could be extended beyond its two-dimension version of sides of a right triangle ($a^2 + b^2 = c^2$) to three dimensions representing the sides and diagonal of a cube ($a^2 + b^2 + c^2 = d^2$), and this could go on to four or five or an arbitrary number of dimensions.

Although it's hard for us to visualize a cube of higher dimension, which is called a hypercube, the same principle can be applied to any shape and extended to n-dimensional space, or hyperspace, as it's referred to today.

To visualize a hypercube, we can use the analogy of a square in two dimensions, which can be extended out into a cube in three dimensions, and extended to a hypercube in more dimensions. It would mean that the world we see is a projection of a four-dimensional (or even n-dimensional) universe into the three dimensions that we can see.

This idea was turned into a highly successful novel by Edwin Abbot, a clergyman in the City of London School, in 1884, called *Flatland: A Romance of Many Dimensions by a Square*. Flatland, meant to be a novel about Victorian politics, caught the imagination of the public. In it, Flatlanders are not allowed to talk about the third dimension, and all laws, including those of Mr. Square, the titular hero, are in two dimensions. Like Plato's philosopher who leaves the shackles of the wall in the cave, Mr. Square realizes there is another dimension and tries to tell his fellow Flatlanders, who are not ready to listen to him.

Prior to Einstein's rise in worldwide fame in the 1920s and the popularization of time as the fourth dimension, this idea of a fourth spatial dimension was quite popular. In his book *Hyperspace*, Kaku profiles Charles Howard Hinton, a British mathematician and science fiction writer, whom Kaku calls "the man who saw the fourth dimension." Hinton, Kaku tells us, even had names for directions in this extra dimension; just like we say left or right, up or down, back and forth, he said that in the fourth dimensions the directions were *ana* or *kata*.[67]

Although the idea of higher dimensions caught on with the public after Riemann introduced the idea in the mid-nineteenth century, it never really caught on with scientists who believed there was no evidence of higher dimensions. Riemann's ideas were thought of as abstract mathematics that had nothing to do with the real world—at least until Einstein came along with his special and general theories of relativity, and Einstein's old mentor, Hermann Minkowski, coined the term *space-time* and time became generally acknowledged as the fourth dimension.

Half a century later, in the second half of the twentieth century, Riemann geometry would serve as the backdrop for other theories of higher dimensions, including string theory.

If there were other physical dimensions around us, it's not hard to see how there could be other things going on, perhaps even entire other universes tucked away in these dimensions, which we can't reach because we are like the Flatlanders in this case. If parallel universes do in fact exist, the toolset of higher dimensions, whether they are spatial or even temporal dimensions, could come in quite useful. Where are they? They are just down there, down *ana* (or *kata*), as the fourth-dimensionalist Hinton would say!

COUNTERPART AND SHARED HISTORY IN A PARALLEL UNIVERSE

In the show *Counterpart*, which premiered in 2017 and lasted only two seasons, we see a two-universe model but one that is more explicit about the relationship between these worlds. We meet Howard Silk (played by J.K. Simmons) who is a quiet office worker working for a secret UN agency in Berlin. At first, we don't know much about this agency, but we discover that it is harboring a big secret. This agency is built over the spot where a physics experiment branched the universes, and there is a tunnel underneath the building that connects the two universes.

Not only is there another version of Howard Silk from the other universe, who is much more suave than the Howard we met (and a spy to boot), there an alternate version of almost everyone. This raises interesting questions in a cold war-like setting, reminiscent of US-Soviet spy exchanges over the Berlin wall, with spies being sent in to replace the other version of themselves. *Counterpart* reveals that the branch happened in 1987, so up until that point, it isn't so much that there were doppelgangers but that there was only one universe. The two Howard Silks before 1987 were the same person; thus, they share the same memories up to the branch. From that point, the universes diverged. In one major divergence, a major plague killed many people in the alternate universe. This makes that universe a strange world, where public spaces are mostly empty, eerily prescient of our pandemic here in 2020–2021 as I write this book.

Counterpart, like much fiction that involves alternate universes, stays away from the science of how the branch happened, but it gets at similar questions of temptation that are common to this genre. If someone died and you knew there was another version of them out there, would you go to see them? If you and your wife (or husband) had divorced and you missed the person that they were, and you found out you were still together in the other universe, would you go in and replace your doppelganger and assume that life? Set against this cold-war backdrop, there are plenty of moral questions as well as questions about how different you might be if you made different choices in the past, embodied by the doppelganger of a particular character.

Chapter 5

Welcome to the Quantum World

Anyone who is not shocked by the quantum theory does not understand it.

– Niels Bohr

Before we can get into the quantum multiverse, let alone the simulated multiverse, we need to introduce some of the key ideas of quantum mechanics and how it relates to the simulation hypothesis. I covered the relationship between the two extensively in my previous book. This was particularly true about the Copenhagen interpretation of quantum mechanics and why it fits the simulation hypothesis much better than it does with a materialist hypothesis of the universe. I'll summarize that briefly later in this chapter, but in this book, I want to go beyond this and explore the relationship between the other popular interpretation of quantum mechanics, the many worlds interpretation (MWI) and the simulation hypothesis, which we will do in the next chapter.

A Quantum of Nature

Before we can understand either interpretation, we have to start with what's called the measurement problem in quantum physics. This is sometimes popularly known as the observer effect or, more formally, as quantum indeterminacy. It gets to the heart of

Bohr's claim that if you aren't shocked by quantum mechanics, you haven't understood it.

Today, most of us have heard of some of the weirdness of quantum mechanics, but it's not something we think about often because we suppose it has very little to do with our daily lives. It seems to apply only when talking about the very small, elementary particles at the subatomic level.

And therein lies the dilemma of quantum mechanics—the rules and principles that were delineated a century ago to explain the behavior of subatomic particles like photons and electronics seem mostly counterintuitive to the macroscopic world that we all inhabit. But if the world we inhabit is built up of atoms, which are built up of subatomic particles, then there must be some relationship between the micro and the macro.

Let's start with a definition:

> **Quantum mechanics:** The branch of mechanics that deals with the mathematical description of the motion and interaction of subatomic particles, incorporating the concepts of quantization of energy, wave-particle duality, the uncertainty principle, and the correspondence principle.[68]

Unlike relativity, for which we can thank Einstein, there is no single founder of quantum mechanics. The honor goes to a cadre of early-twentieth-century scientists who kept finding surprising things as we started to understand the atom and what happens inside it.

The term *quanta* (plural for quantum) was used well before quantum mechanics came to be; it was a term used to specify the smallest discrete amount of something. Yet the discovery that certain quantities in nature exist only at discrete values was surprising. Newton's laws of physics and science before the twentieth century assumed that the world was continuous, which made mathematics the ideal tool to represent it.

As an example, there could be an infinite set of numbers between 1 and 2. On a planetary scale, you could think of orbits of asteroids and planets. Even though Earth and Mars are separated by millions of miles, there is no reason you couldn't have an asteroid at, say, 1.33 AUs, or 1.34 AUs, or 1.35 AUs (an AU is the distance of the earth from the sun).

But what if you could only have asteroids at 1.2, 1.6, or 2.0 AUs? If you tried to place an asteroid anywhere in between (like at 1.33 AUs), what would happen? What if it magically jumped to either 1.2 AU or 1.6 AU? This would be surprising indeed! In this instance the quantum of the orbit (the smallest discrete value allowed) would be .4 AUs.

This is what happened when Niels Bohr, the Danish scientist who will factor heavily in the interpretations of quantum mechanics, put together the model of the atom that is still embedded in most schoolchildren's brains today: the planetary model. It is a classical model in which the electrons circle the nucleus like the planets of a solar system. When Bohr was trying to figure out the structure of an atom, he had a dream one day about a horse race. The horses were running around the track in lanes that were divided by white chalk. When a horse tried to cross certain tracks, it wasn't allowed.

Using this dream as intuition, Bohr realized that there were discrete values, or energy states, that electrons could have, and that it wouldn't transition gradually from an energy state such as 1.2 to 2.0.

Quantum mechanics isn't about orbits, really, but about quantization, the idea that quantities like energy and momentum only take on discrete values rather than the more continuous behavior seen in classical physics systems.

The way that electrons move up is by gradually adding energy to the electron, and when some threshold is reached, it would execute the jump to the higher energy state (called a quantum leap).

Similarly, if energy was removed, it would jump back down to the lower orbit and release some energy at some point. Electrons were not allowed anywhere in between these discrete energy states, which can be thought of as fixed orbits, just like our example of asteroids around the sun.

Bohr wasn't the only one or even the first one to find this strange behavior when looking at the states of energy. Bohr proposed his theory in 1915 and was awarded a Nobel prize in 1922. In 1900, Max Planck was studying black body radiation, radiation emitted by a black body that is opaque and not reflective. Planck found that energy could be emitted not continuously but only discretely. Planck called it a quantum of energy and laid the groundwork for all of the quantum theory.

Although Einstein wasn't a big fan of certain aspects of quantum mechanics, his only Nobel prize came not for his theories of relativity but for his work with the photoelectric effect, which he published back in 1905. In it, he states:[69]

> Energy, during the propagation of a ray of light, is not continuously distributed over steadily increasing spaces, but it consists of a finite number of energy quanta localized at points in space, moving without dividing and capable of being absorbed or generated only as entities.

This quantization of light that Einstein referred to is what we call a photon today. This behavior of light as a photon, which acts like a particle, was puzzling, because light was thought to be a wave. In fact, Maxwell's equations, formed by Scottish scientist James Clerk Maxwell in 1861 and 1862 for the study of electromagnetic phenomenon, assumed that light was a wave.

How could it be that light and electricity, both of which were thought to be waves, were also quantized (i.e., acted like discrete particles)?

Quantum Mechanics: The Wave and the Particle Duality

This duality between a wave and a particle brings us to the experiment that is often at the heart of quantum mechanics: the double-slit experiment, a version of which is shown in Figure 7.

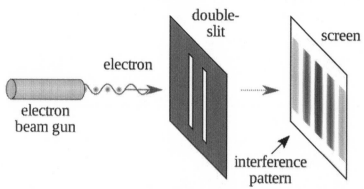

Figure 7: The double-slit experiment that shows that electrons behave as waves or as particle. [70]

The basic concern is that if you have a screen behind an object with two slits and shine a light through the silts, what happens? Normally, there is an interference pattern that appears on the screen beyond. The interference pattern indicates that light is displaying the behavior of a wave, going through both of the slits and creating the pattern on the screen. On the other hand, if light consisted of particles, or photons, then you should see two clusters of them on the screen across from the slits. The puzzling thing is that when individual particles were observed near each slit, the behavior of light reverted to individual-particle behavior, with only two clusters on the screen, across from slit A and slit B.

The experiment had been done with light back in the 1800s, but it was a confirmation of the experiment using electrons in 1927 that showed that this behavior was more than just an oddball property of light. If electrons displayed similar behavior to light, was it possible that other subatomic particles also displayed this same strange behavior? In fact, experiments have been done since that show that even entities as large as an atom might display the same strange behavior predicted by quantum mechanics, known as the particle-wave duality.

Schrödinger's Wave Function

Electrons in particular were found to behave strangely. Not only could they exist only at certain energy states (quantized), but also that it is difficult to know the precise location of an electron until you actually measure it. This is why physicists now refer to an electron cloud, which is a cloud of the possible locations of the electron.

Erwin Schrödinger, an Austrian-Irish scientist, was also one of the fathers of quantum mechanics. In 1926, at the age of 39, while at the University of Zurich, Schrödinger came up with an equation to define the behavior of these particles, called the Schrödinger wave equation. This equation provided the mathematical underpinnings of the newly developing quantum theory (building on the earlier work of Planck and Bohr and Einstein), describing the wave of probabilities that went along with particles. The Schrödinger equation (as it's called for short) is considered so important in the newly developing quantum theory that it "bears the same relation to the mechanics of the atom as Newton's equations of motion bear to planetary astronomy."[71]

The wave equation, when it was solved for a particular particle, output a wave function that showed the probabilities of different physical events, such as the location of a particle. Because it was a wave function, it encapsulated all the probabilities of all the places a particle could be. It applied only to a single particle and not to all particles in the universe.

According to quantum theory, a particle doesn't have a definite position, at least until it's measured; it has only a set of probabilities that show where it *might* be.

A good way to think about this is to imagine going to a movie theater in which you know there is one particular patron, but you don't know which seat that patron is in. Schrödinger's wave equation shows the different possible seats that the patron might be in, and the amplitude of the wave above each seat in the theater represents the probability that the patron is in that particular seat, as shown in Figure 8.[72]

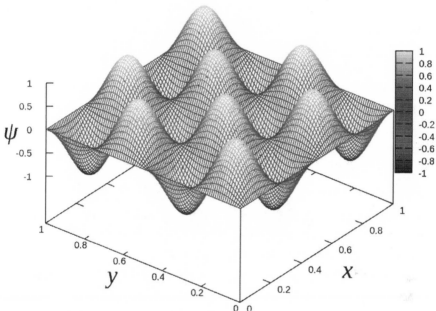

Figure 8: A graphical representation of a probability wave[73]

The Collapse of the Wave in Copenhagen

This was one of the first introductions of probability into physics, which threw many eminent scientists for a loop and began to challenge the centuries-old notion that we lived in a purely deterministic universe, which has held since the time of Newton and French scholar Pierre-Simon Laplace.

In 1814, Laplace famously articulated a thought experiment (now called Laplace's demon):

> If there was an entity (the demon) that knew all of
> the initial conditions (precise location and
> momentum) of every particle in the universe, the
> demon (or a machine) could predict exactly where
> every single particle in the universe would be at a
> certain time in the future or the past, using the laws
> of classical mechanics. Laplace's demon relied on
> being in a purely deterministic (no randomness) and
> completely reversible universe.

By introducing probabilities into the location of particles in the universe, the quantum theory was upsetting not only Laplace's demon but everyone who believes that we live in a completely deterministic universe. Even Einstein didn't like this idea, leading to his famous objection to the quantum theory, that "God does not play dice with the universe." Schrödinger himself was upset by this interpretation of probabilities, even though his equation did in fact make accurate predictions.

Even as the founders of this new theory struggled to come up with an explanation that would help in understanding what was going on, the undeniable facts were that when you measured a particle, it would be at one definite location. And if you didn't, it was in a bunch of possible locations.

This resulted in the Copenhagen interpretation of quantum mechanics, which stated that although a particle's location may exist as a wave of probabilities, once it is observed (or to be more precise, measured) it took on a definite location. This was called by Niels Bohr (who was located in Copenhagen, Denmark) and his colleagues the "collapse of the probability wave."

Going back to our example of the patron in the theater, it's as if you walked into the dark theater and pointed a flashlight at a single seat, you could tell if the patron was there or not. At that moment, the probability wave is said to have collapsed into a definite reality where the person is or isn't at that location, so the probabilities of the other seats in the theater no longer need apply.

There is an ongoing debate about what constitutes observation.

Is it a simple measurement device, or does it require a conscious observer? One argument is that even a measuring device that is not a conscious observer is enough to collapse the probability wave, so a conscious observer is not needed. Another argument goes that even if the measurement device records the subject, there is still the probability of *multiple different* values being recorded on the device, and it's not until a conscious entity observes the measurement that we can be certain it really happened that way.

This taps into a broader discussion not just of multiple presents but of multiple pasts, which introduces a whole new level of weirdness about time and multiple timelines that we'll talk more about in the next chapter.

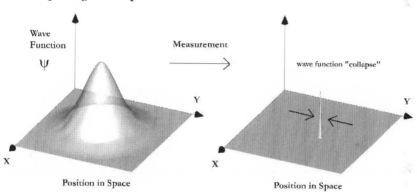

Figure 9: An example of the wave function collapse when observed.[74]

Superposition and That Ridiculous Cat

Since everyone can agree that particles, after they are measured, take on a definite position, where is the particle before measurement is done?

The particle is said to be in a state of *superposition*, meaning a superset of all positions. For example, an electron could be spinning upward or downward. The term was derived from Schrödinger's equation, because you could add up two or more quantum states and they would add up to a valid quantum state that was a superset of its constituent states. The Schrödinger equation then showed a superposition of all possible states of a particle.

Today when we refer to a particle (or a qubit, a quantum bit) being in superposition, we mean an object that hasn't yet taken on a definite value. Another way to think about it is that it has multiple possible values at the same time.

If this seems absurd to you, you aren't alone. Even Schrödinger himself didn't like the idea of an object being in more than one state at a time (i.e., in superposition). He invented the now famous metaphor of Schrödinger's cat to show what a ridiculous idea it was. Ironically, it has become perhaps the simplest way to understand quantum superposition, which has become the basis for quantum computing.

In this fictional example, Schrödinger places a cat in a box along with a vial of poison and a single atom of radioactive material. After an hour, the atom has a 50% chance of releasing radioactivity that would unleash the vial of poison, which would kill the cat. Obviously, then, the cat has a 50% of being alive and a 50% chance of being dead after an hour.

So, after an hour, is the cat alive or dead? Common sense tells us that the cat is alive or dead, it can't be both. Quantum mechanics seem to imply that until a measurement or observation is made, there is no definite state: The cat is said to be *both* alive and dead— that is, it is in a superposition of those two states. Schrödinger himself created the example to point out that this makes no common sense whatsoever.

The First Wall: Object versus Subject

Actually, at least two major issues cropped up with the new quantum theory that made many scientists uncomfortable, which I call "Two Walls of Quantum Mechanics." We've already mentioned that Einstein and Schrödinger in particular were not fans of the need for probabilities, as evidenced by both Einstein's famous statement about God not playing dice and Schrödinger's cat argument to show how nonsensical the idea of an object in superposition was.

Moreover, the collapse of the probability wave implies the breaking of a Chinese wall that scientists try to impose on the universe as they study it: that of observer and subject. John Wheeler in particular contributed a fascinating insight (out of his many) about this first wall.[75] Wheeler concluded that the universe couldn't be studied in the way that scientists had been formulating, with the scientists being impartial observers watching what happened in the experiment.

Rather, Wheeler concluded that we lived in a "participatory universe" and that there was no such thing as objectivity, because the scientists who must be there to observe the experiment are in fact part of the experiment. I quoted Wheeler in *The Simulation Hypothesis* and, at the risk of repeating myself:

> Nothing is more important about the quantum principle than this, that it destroys the concept of the world as "sitting out there", with the observer safely separate from it by a 20-centimeter slab of plate glass. Even to observe so minuscule an object as an electron, he must shatter the glass. He must reach in ... to describe what has happened, one has to cross out the word "observer" and put in its place the new word "participator." In some strange sense the universe is a participatory universe.[76]

That the scientist is participating in the experiment is almost scientific heresy from the point of view of classic science. Wheeler is saying that the first wall of science—that between objective reality and subjective reality—doesn't really exist. Or at least that it is malleable. Moreover, it gets into an uncomfortable space that the materialist point of view tries to avoid at all costs: defining subjectivity and consciousness, or acknowledging that it might somehow have an impact on the physical world.

Max Planck himself said that he believed it would be impossible for scientists to understand the material world totally, because we are part of that world: "Science cannot solve the ultimate mystery of Nature. And it is because in the last analysis we ourselves are part of the mystery we are trying to solve,"[77] an almost direct echo of Wheeler's later statements about a participatory universe.

But many physicists, then and now, still cling to the classic point of view that there is an objective physical reality and that although we live in it, we do not and cannot, with our observations or other subjective mental experiences, influence this world in any way. As such, any measurement device will do to collapse the probability wave, argued many physicists, and there was nothing special about consciousness. Einstein, ever a critic of the new quantum theory, didn't like the idea of an observer being necessary, and only half-jokingly wondered whether a sidelong glance from a mouse would count as an observation.

Whether you define these interpretations of quantum mechanics as requiring a measurement device (which is part of the system being observed) or requiring a consciousness (to observe either the original collapse or to observe the measurement in the future), you can see the importance of this wall being erected or demolished, depending on how you choose to define the wall. The uneasiness with this wall being broken caused scientists to look askance at to look for an alternative to the Copenhagen interpretation.

The Second Wall: Small versus Large

Yet, in the years since its first formulation, quantum mechanics has been shown to be experimentally correct, displaying distributions of particles as predicted by the quantum equations in pretty much every experiment conducted to date.

To explain why large objects like books and tables weren't

popping into and out of existence or displaying probabilistic behavior, Bohr drew a metaphorical line in the sand that said these rules only apply to the very small.

Meanwhile, Einstein's general theory of relativity was successful in its predictions of gravity's effect on light, and even Einstein's special theory of relativity's predictions about time dilation were shown to be correct experimentally.

This puts scientists in a strange position, with one theory that works for very small objects (quantum mechanics) and one that works for very large objects (general relativity), but are separated by a wall that seems artificially put in place by the proponents of the Copenhagen interpretation.

The main objection to having such a wall is that the universe as we know it is made up of these small particles, so there must be some impact of this strange behavior on the whole.

In fact, quantum theory was only applied to subatomic particles for many years, but the more experiments that are done, the more suggestive it has been that quantum mechanics does apply to larger objects. Today, objects as large as atoms and molecules have been shown to exhibit quantum behavior. If you talk to physicists today, they say that it's not that quantum mechanics doesn't apply to larger objects; it's that the probability waves look different.

Greene says that there is "every reason to believe that it works for collections as hefty as those making up you and me and everything else." [78] Greene goes on to explain that there may be no such wall, because the probability wave becomes less distributed with larger objects.[79] If you throw a baseball, for example, it's likely to land where Newton's laws of motion say it will. But "likely" is not the same as 100%; it means that there is perhaps a 99.9999% chance that it will land where Newton's equations predict, and a .00001% chance it will go somewhere else, like landing outside the stadium.

These two walls—between object and subject and between large and small—led physicists to realize that quantum mechanics is showing us a very different picture of the world than classic physics or even Einstein's relativistic physics, and to look for alternative explanations to the Copenhagen interpretation of a collapse of the probability wave, which leads us to the alternative: a multiverse interpretation of quantum mechanics, which we'll delve into in the next chapter in detail.

Simulation and Quantum Mechanics

One of the arguments I made in my previous book is that many of the baffling findings of quantum mechanics are only baffling if we strongly hold a materialistic view of the universe. On the other hand, if we adopt an information-centric model of the universe, which the simulation hypothesis presupposes, these things don't seem so strange at all. I'd like to give a quick high-level summary of some of these arguments, though you should read the previous book to get more details.

Pixels and Quanta

Today's physicists generally acknowledge Planck length (1.6×10^{-35}) as being the smallest amount of space in which anything can be measured. This suggests a type of pixelization where the pixel is the smallest addressable unit of space. Note that even on a computer screen with, say, n pixels across, you can still do math that involves, say, a half or a third of a pixel, but those computations can't be translated into the physical screen on which the universe is rendered. Similarly, I can write an equation that involves $1/10$ of the Planck length, I just can't measure that amount in the real physical world.

This seems to be consistent with the quanta in the quantum theory, that certain things—including energies of states that a particle can exist in—can only be certain discrete values. Newton's equations assumed a continuous amount of space; it turns out the universe may be more quantized than we thought, suggesting a type of pixelization of our universe.

In software, you can always calculate smaller values abstractly, but you can't actually measure or draw anything smaller than a pixel when you are rendering it on a screen. It's as though, if you want a pixel to light up, you can keep adding up continuous numbers until a threshold is reached and then the pixel lights up.

Is time also quantized? This is an open debate. Some have defined the Planck time, the amount of time that light takes to traverse the Planck distance, as the smallest measurable unit of time. There are physicists who think that for the quantum theory to be unified with Einstein's general theory of relativity, a quantized time would be quite useful. William G. Tifft, a professor of astronomy at the University of Arizona, has suggested that there may be quantization in space and time.[80] This would be just like a computer with a specific clockspeed, which can only measure intervals that are some multiple of this speed.

There Is No Such Thing as Matter, Just Information.

Most of us know that an object as solid as a table is made up mostly of empty space if you zoom in. In fact, the atoms themselves are mostly empty space, consisting of subatomic particles like electrons, protons, and neutrons. But what about these subatomic particles? The more that science peers in, the more strangeness it finds in the subatomic world; particles appear as probability waves, with a certain probability that that particle exists at that point in time. In fact, the whole idea of "particle" may not be what we think of it at all, but rather that it is a set of information.

John Wheeler, one of the giants of twentieth-century physics, said that in his lifetime, physics went through three phases. In the first phase, we assumed that everything was a solid particle. This corresponded to classic physics. In the second phase, we thought everything was a field because of the emergence of the quantum theory. In the third phase, he surmised that everything was actually information. Everything we thought of as a particle was really a set of answers to a number of yes/no questions, or bits, as we think of it in the computer science world. Wheeler coined the famous phrase, "it from bit," which reflected his belief that everything that is real actually comes from bits. Oxford's David Deutsch modified it to reflect the idea that every quantum particle is really a bit, as in a quantum computer, which we'll talk more about in Chapter 10: Quantum Computing and Quantum Parallelism.

The Collapse of the Probability Wave, Quantum Indeterminacy

As we mentioned earlier, the interpretation of the collapse of the probability wave is one that is befuddling to most scientists and to our common sense. The observer effect has generated much debate and ties in with our ongoing discussion of NPC versus RPG modes of a simulated universe. Does a conscious observer need to exist for the probability wave to collapse, or does it happen if there is any kind of measurement? Some, like Planck, Eugene Wigner and John von Neumann, some of the greatest minds of twentieth-century physics, believed that consciousness was needed, whereas others try to avoid this possibility. Planck wrote: "I consider consciousness as fundamental and matter as derivative."

What no one has been able to figure out is *why* quantum indeterminacy exists.

I and many others believe the simulation hypothesis provides a pretty good answer. The reason that video games have advanced so far in a few decades is because of optimization techniques. It would be impossible even for today's computers to render in real time all the pixels of a single 3D world; instead, information is stored as 3D models outside the rendered world and then only what a particular character can see from a certain angle is rendered. In short, only that which is being observed is rendered.

Many adherents of the simulation hypothesis think that quantum indeterminacy is simply an optimization technique with the same basic idea: only render that which is being observed so that not every particle in the whole universe has to be rendered at one time, only those which are being observed. Everything else is in a state of superposition, or stored simply as information. If there's one thought I want to leave you with about computer science and information theory, it's that optimization of information is one of the key ways in which we accomplish seemingly impossible things. A more detailed overview of both quantum indeterminacy and quantum entanglement as optimization techniques is given in *The Simulation Hypothesis*.

Future Selves and Parallel Universes

Quantum mechanics is telling us weird things about reality, and even about time. In the first instance, it is implying that there are probabilities and that we might exist in multiple states of superposition. In another, time is very strange in quantum mechanics, and the future and the past take on different meanings than we have in ordinary life. We'll explore these two aspects in the next two chapters.

Chapter 6

The Quantum Multiverse

We are now ready to get back to our list of multiverses and dig into the one that is most relevant to the simulated multiverse: the quantum multiverse. In the previous chapter, I described how some scientists were uncomfortable with the Copenhagen interpretation, even if they couldn't quite put their finger on why.

There are, of course, many other interpretations of quantum mechanics, including the von Neumann-Wigner variation on the Copenhagen interpretation that required a conscious observer, and David Bohm's interpretation, called Bohmian mechanics (or more formally, de Broglie-Bohm theory). However, the one that has become the most popular alternative to the Copenhagen interpretation is the many worlds interpretation.

Type 5: The Many Worlds Interpretation

Ironically, Wheeler, an advocate for the participatory universe, would play a big role in this alternative interpretation. In 1957, Wheeler's student at Princeton, Hugh Everett III, proposed an alternative to the collapse of the probability wave as part of his PhD dissertation. He suggested that the wave function, rather than collapsing, separated into different wave functions, one for each possible measurement. Although the mathematics worked, the meaning of such decoherence, as it's referred to, was that if you split the wave into all its separate components, each one existed by itself in a separate physical universe (i.e., without the others).

This would mean that there were literally different physical universes that were created whenever a quantum choice was to be made. What started out as an obscure thesis has come to be known in time as the MWI and is now championed by many physicists. How that happened is an interesting story.

Wheeler, who originally championed the MWI of Everett and agreed with his math, actually discouraged Everett from insisting that his math implied there were many *physical* worlds. This was because both Bohr and Einstein had issues with the interpretation, if not with the math.

Bohr and his colleagues objected because they preferred the Copenhagen interpretation of the collapse of the probability wave. They were also uncomfortable with the implications of the theory that many physical worlds were being created, which one colleague of Bohr's said was like "theology."[81] There is, no doubt, a magical element to the idea that many physical universes are being created in every instant (unless of course, we are talking about a simulated multiverse, but we will get to that later in the chapter).

When Wheeler took the idea to Einstein, who still had an office in Princeton, Einstein objected not so much to any specifics of the interpretation, but because he still didn't believe, in his heart, despite the confirmation by many experiments, that quantum theory (either a probability approach *or* a many-worlds approach) was how the universe really worked.

For this reason, Wheeler, not wanting to offend either of these two legends, asked Everett to pull back on the nomenclature and the idea of many distinct physical worlds. The thesis went through many titles, ranging from *Quantum Mechanics by the Method of the Universal Wave Function* to *Wave Mechanics Without Probability*, to a shorter version titled, *On the Foundations of Quantum Mechanics* that had some chapters removed. This was the one that Everett successfully defended to get his PhD in 1957.[82]

Wheeler's thesis relied on the idea of *quantum decoherence*, which means that particles that were previously entangled or connected could become disconnected and go their separate ways. In this case, a particle in superposition would split into two versions of itself, one with the wave function of one alternative being realized, and the other with the wave function of another alternative, and so on for each alternative. In this way, you can think of decoherence as the opposite of superposition, which aggregates all the possibilities; decoherence pulls out each possibility individually.

Everett, who was eager to get his doctoral degree, agreed to pull back on language implying the creation of many worlds and to stick to the math of the wave function. His dissertation was approved and Everett moved on to his new job in private industry. As a result, the dissertation sat in obscurity for years.

After a number of years, Bryce DeWitt, who had given some critical comments initially when he was asked to be one of the original reviewers of Everett's paper, came around. Interestingly, by this time, Wheeler had somewhat cooled on the idea of multiple worlds. In 1970, DeWitt published a paper in *Physics Today* about Everett's thesis and further championed the idea by publishing Everett's paper in book form, called *The Many Worlds Interpretation of Quantum Mechanics*. In fact, it was DeWitt, not Everett, who coined the phrase, "many worlds interpretation," which is still used today.

Starting with a story in Analog *magazine* in 1976, this has led to a rash of parallel-universe stories.[83] Eventually, the term *multiverse* caught on, or, more specifically, *the quantum multiverse*, a term used by Greene and others.

SLIDERS AND WORMHOLES

In *Sliders,* we meet physics student Quinn, who is working on generated wormholes (a popular way to traverse between universes, at least in sci fi), when he is visited by a version of himself from another universe. The other Quinn shows our hero how to make a device to slide between universes, and Quinn is joined by his professor, Maximillian Arturo (played by the ever-present John Rhys Davies, who has appeared in everything from the *Indiana Jones* films to *The Lord of the Rings*) and friends. Their group slides into different parallel worlds, and the premise of the series is that they don't quite know how to get home. All of the worlds deviate a little (or a lot) from our own world, which is called the prime universe. This includes a world where gender roles are reversed, where Hillary Clinton is president (remember this series ran before she actually ran for president), and men are only allowed to have domestic jobs; the Soviet world, where the Soviet Union won the Cold War; Tundra world, where a nuclear winter has caused a deep freeze; and last but not least, Elvis world (don't ask me to explain this one).

A Tree of Universes or a Single Big Multiverse?

Everett was actually originally discussing what would happen if you made repeated measurements of a particle, in which case he suggested that each measurement caused decoherence of the wave function so that it split into multiple wave functions. As a result, if you continued to make measurements on a particular particle, you would continue to see this splitting, as illustrated in Figure 10.

This would mean that every time a quantum decision is made, this decoherence happens in an entirely deterministic way, creating a new version of the particle in each reality *every single time there is a measurement.*

If we assume that each decision is a binary decision, we now get the kind of tree shown here and which we have been discussing in this book: one world, branching off into many worlds, each, in effect, a timeline that evolves separately from the others (which we might call its *sibling* timelines).

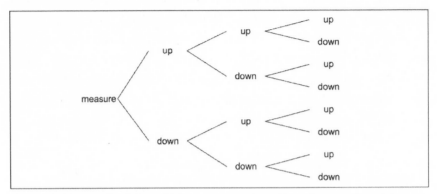

Figure 10: An example of the multiple measurements and how they would result in many worlds.

What Everett was implying was that there was a gigantic wave function that contains within it all possible realities, that is, all superpositions of the particles that are represented. Everett referred to this as the *universal wave function*. Philip Ball, writing for *Quanta Magazine* gives a succinct description of what is happening:

> ...[A]s it (the universal wave function) evolves, some of these superpositions break down, making certain realities distinct and isolated from one another ... [W]e should speak of the unraveling of two realities that were previously just possible futures of a single reality.[84]

This means that what we are seeing as the tree grows are possible futures, each of which becomes present realities when the measurement is made. This is why the idea of multiple timelines and the MWI are intimately linked.

Schrödinger and alternate histories

It turns out that Everett wasn't the very first physicist to bring up this idea, though he was the first to really formalize it. In fact, Schrödinger himself, not a fan of the collapse interpretation, had brought up the idea in a lecture in Dublin in 1952, while allowing that the idea might "seem lunatic." Rather than using parallel worlds, Schrödinger described several histories, explaining that they were "not alternatives but all really happen simultaneously."[85] This statement is quite shocking when you think about it, not just from the point of view of multiple histories but of our normal understanding of time and causality, which we will explore in the next chapter. What exactly does "multiple histories" that are "happening simultaneously" mean if not multiple timelines?

You can see that these ideas are all related: multiple timelines, the different histories of Schrödinger, and the parallel universes of the quantum multiverse.

Neither Schrödinger's original speculations nor Everett's formalization of the MWI was accepted at first. But as the years passed, and no successful alternative to the Copenhagen interpretation was found, more physicists have added their names to the interpretation as being most likely. Make no mistake: neither of these two popular interpretations has been proven to be the definite way things happen at this date. In fact, it may not be possible to prove any of them, which is why they are referred to as most popular *interpretations* of quantum indeterminacy.

How the Quantum Multiverse Deals with the Two Walls

As far as I can tell, one reason many physicists have started to prefer this interpretation over the Copenhagen interpretation because it doesn't require this strange thing called a collapse of a probability wave, nor does it require another strange thing called an observer.

In a sense, it does away with both walls that we discussed earlier. On the one hand, it doesn't have to acknowledge that there is anything called consciousness, even though many physicists today insist that consciousness isn't necessary for the collapse of the probability wave, just a measurement.

On the other hand, MWI also does away with the wall between the very small and the very large that was put up by Bohr and others when restricting quantum mechanics to the realm of subatomic particles. As mentioned earlier, subsequent experiments confirm that quantum effects do seem to happen with bigger objects, and many physicists insist that baseballs and other objects also have quantum effects; we just don't observe them very often.

Although the idea of a baseball disappearing or appearing somewhere else sounds completely bizarre, the MWI gives an out. The other possibilities are happening; you just can't see them because they are in *another* universe. This consequence of the MWI means that you don't need separate theories for the very big or the very small.

The quantum multiverse, because it is becoming accepted by a large number of physicists as being a very real implication of the equations of quantum mechanics (which have been shown again and again to be accurate), is the kind of multiverse that we are most interested in.

What Happens to the Wave?

Before we move forward with the quantum multiverse and look at how it relates to the simulation hypothesis, let's review some important aspects of this multiverse, starting with the wave.

Both the Copenhagen interpretation and the MWI agree that there is a wave for the different states of the particle. In the Copenhagen interpretation, there is a collapse of the wave, and it transforms from the left side of Figure 9 to the right side: from many possibilities to a single possibility. Physicists remind us that there is no math that gets you from A to B, because we don't fully understand the collapse of the wave, though pretty much everyone agrees that when some measurement is done, only one possibility is left, and it looks like the right side of the figure. In that sense, it relies on a kind of magic called observation or measurement.

Everett's MWI, on the other hand, takes advantage of the mathematical properties of a complex wave function. Basically, the reason that the results of the double-slit experiment suggested a wave was because the interference patterns seen on the screen match those of waves—specifically, we see patterns called constructive and destructive interference, where you can add or subtract waves from one another to get a complex wave that is, in a sense, the combination of those waves, as shown in Figure 11.

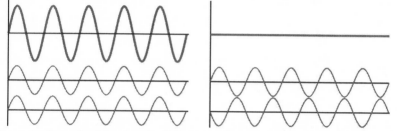

Figure 11: Example of Constructive (left side) and Destructive Interference (right side)[86]

In the Everett/DeWitt view of the world, the complex wave is exactly that—a combination of waves. It is, in effect, the sum result, the superposition of individual waves, which look like the lines on the top of Figure 11 (the resultant wave is on top, consisting of the two waves below it). In fact, although we think of the two waves being combined to produce the resultant wave, you can go the other way and start with the top resultant wave and cause decoherence, or splitting back into its individual wave functions.

If we start with the complex wave, how do we end up with just a simple wave? What happens to the other interfering waves? Practically speaking, this means that all the other possibilities disappear somewhere. Where? That's where the idea of other parallel universes, physical or otherwise, comes from, since the information in the wave isn't allowed to be destroyed.

In effect, the probability wave, then, is a superposition wave of all possible states of the particle across all possible universes, each of which is interfering with the others. You can see why interference is actually a pretty important part of the quantum multiverse theory. It's built into the Schrödinger wave equation and must be accounted for in any theory.

Multiple Universes: Physical or Probable or Digital?

One big fundamental question that comes up when discussing MWI is whether these different universes are real in the same sense that we think our physical universe is real. Everett sensed an upcoming debate that would crop up because of his theory. In the old interpretation, although there were many possibilities, they weren't all realized; only that which is measured was realized (or as computer scientists might say it, rendered).

This debate, about whether all of the universes that are implied by the MWI are actual physical universes or just probabilities is one that has never been fully resolved. Everett went out of his way to say that they were actual physical universes, though there was no way to prove this. The implication was that we were each being cloned every time there was a random decision—a quantum coin toss. In fact, the whole universe, in a sense, is cloned every time a quantum measurement is made in this interpretation.

Max Tegmark tells of meeting Bryce DeWitt, the main force behind popularizing Everett's theory in 1970, who tells of meeting Everett. De Witt told Tegmark that although he liked Everett's math, he complained to Everett that he "didn't feel like he was being cloned." Everett then asked him if he "felt" like he was orbiting the sun at some thirty kilometers per second, to which De Witt replied no. After this, DeWitt realized that was no reason to object to the many-universe theory.[87]

Fine-Tuning, Revisited

Let's now revisit fine-tuning, which seems to make sense if there are multiple universes, each with different variations of fundamental constants. Ours just happens to be the one that allows galaxies and stars and carbon-based life. This would, of course, imply that there are a very large number of universes, though part of my thesis in this book is that the universes spread in a tree that is then trimmed or pruned, which reduces the number of universes that exist at any given time.

One suggested solution is that someone or something (a designer) tuned the variables in setting up the initial conditions, and then like a computer simulation that proceeds with deterministic rules, our universe was allowed to evolve to the current point.

Even if the initial conditions were varied randomly across multiple simulations, only those that produced life might be interesting to the simulators, and the others might be shut down; thus, we have at least one mechanism in a simulated multiverse for pruning that is consistent with how we run simulations. We'll explore the ideas of the Multiverse Graph and Core Loop, of how this might work, in Part IV.

Quantum Cloning and the Multiverse

The idea of cloning oneself seems silly in a physical sense: it might now seem like quantum physicists are now reverting to magic yet again. Although mathematically there is no problem with decoherence, is there anything in nature that clones itself instantly?

You may have seen recent news reports that particles were teleported from one location to another, using a process of quantum teleportation. This sounds quite esoteric, kind of like the teleporters that are used on *Star Trek* to send Captain Kirk and Mr. Spock to the surface of a planet instantaneously from the *Enterprise*, which usually stays in orbit.

Quantum teleportation relies on a type of entanglement and doesn't actually clone a particle or even, for that matter, teleport a particle. What it does is clone (and if you like, teleport) the information that is contained in a particle. Whereas we can clone the information in a particle by using quantum entanglement with another particle, there is nothing that produces new particles that are exactly the same as the old particle per se. Rather, it is like applying the information from one part of the video game to pixels in another part of the simulated universe.

Nothing in nature clones large objects instantaneously. As we are discovering, even genetic clones of living things have to be grown from single cells and take some time. Biological reproduction takes nine months to create a new mini-clone of a human, a baby, and even then it's not a perfect clone. In general, creating physical things requires some amount of processing time because nature relies on algorithms.

In the MWI, you're not just cloning things once or twice; you are cloning the entire universe every single time there is a quantum decision. That means you are cloning on the order of 10^{80} particles every single nanosecond (or whatever the interval is between quantum measurements).

Further complicating things, do these multiple worlds have *persistence*? In other words, if there are two possible measurements of a particle, does each world branch off and continue to exist ad infinitum? Or does it, with the next measurement of a particle, merge with some other universe that previously had differences? Do some of the universes disappear if nothing interesting is going on, or do they always grow toward infinity?

Although there seem to be massive challenges with cloning physical things, we don't need a long process to clone information. In fact, this is where Wheeler's later insight, that the universe consists of information ("it from bit"), seems to tie surprisingly in with the simulation hypothesis *and* to the quantum multiverse. It is much easier to think of cloning the information of a universe than it is to think of cloning the actual universe.

This is yet another example of information theory seeping into physics. The conservation of energy principle, one of the bedrocks in physics, has been augmented by the conservation of information principle in the newly emerging field of digital physics.

Representing a Quantum Multiverse Digitally

We'll look more closely at how to represent worlds digitally in the next two parts of this book, but one way to think about the multiverse is that it is a collection of nodes; each node is a particular configuration of *all* the particles in that universe, which can be represented by a (very large) binary string.

If we had on the order of 10^{80} particles in the physical universe, and we had 10 bits that represented the answers to the yes-no question that Wheeler postulated in his it from bit, we would need 10^{81} bits to represent the information digitally in the universe (10 bits per particle multiplied by 10^{80} particles).

If we think of a network of nodes, with each node representing one of all the possible values of these bits, then the resulting network pretty much represents the multiverse, an idea we'll pick up with the Multiverse Graph in Part IV.

When the universe branches, all it does is move to a node that has a difference of that one bit. In this model, the information that is needed is really just the difference between each node. If we start with all the possible nodes in the universe as all the possible values of the bits, we don't need to copy information continually (which could result in a very large amount of information). In fact, we wouldn't even need each node to be the same size, since they are exactly the same as the previous nodes with a one bit difference.

Whereas physicists like to rely on the magic of infinity, digital physicists and computer scientists hate having an infinite amount of anything, even information. Much of computer science is about how best to optimize the storage of information and the algorithmic processing of information.

The fascinating thing about this node framework is that it would work even if these parallel universes weren't universes at all, but were instead different theoretical timelines, which could be run as needed. In this case, a timeline would be a particular collection of nodes. But since one node is not much different from its neighbor (only off by one bit), we can reduce the information requirements significantly (not unlike, say, compressing an image of a wintry scene in *Game of Thrones* by recognizing that most of the pixels are simply white). This will be the essence of the Multiverse Graph, an efficient way to represent the multiverse digitally.

The Simulation Hypothesis and the Quantum Multiverse

Let's close this chapter by tackling the simulated multiverse head-on: I have argued that a digital multiverse is much more likely than a physical multiverse for reasons of information storage, compression, cloning, and processing. In quantum mechanics, the wave represents either probability of certain things happening if we run the process multiple times. In fact, that is kind of the definition of probability: that if you do something multiple times, these are the possible futures.

The only mechanism we know of that allows us to run the universe multiple times would be a simulated universe. If run on a type of computer, would allow us to run multiple scenarios either serially or in parallel. In this model, a single timeline then would a be a single run of the simulation, and what we call the present versus the past or the future would be simply a particular node in this theoretical (admittedly giant but not infinite) network of nodes.

In my previous book, I argued that the simulation hypothesis explains both the underlying mechanism as well as the big-picture questions related to the Copenhagen interpretation. At this time, I'm arguing that the simulation hypothesis is a better explanation of the MWI (again, much better than the materialist hypothesis).

But it gets better. The simulated multiverse allows us to get away from having to choose *either* the Copenhagen interpretation or the MWI. You can have both, and the universe we see is the one that is represented as information being run (or rendered) at the present moment in time, as needed, and able to navigate anywhere in this very large set of nodes.

This is, of course, a big leap from the point of view of established physics, but in this book we are asking two questions: *What if?* and *How would it work?*

In our case, a simulated multiverse, based on information, gives us a much better and more understandable framework to think about multiple timelines operating in parallel than the materialist hypothesis does.

However, to talk about timelines more formally, we have to try to answer a strange question that seems simple: What is time? That's not an easy question to answer, and we'll see that the same quantum experiments that gave us quantum weirdness gets weirder even yet when we think about the past and the future, perhaps putting the final nail in the coffin of traditional ideas of both space and time.

CRISIS ON INFINITE EARTHS IN THE ARROWVERSE

Crisis on Infinite Earths was a crossover TV event across a number of TV shows (collectively called *The Arrowverse*) on the CW network in 2019 based on DC comics. It is perhaps one of the best examples of the quantum multiverse in science fiction on TV. The Arrowverse is a multiverse that is named after the first of these superhero shows, *The Arrow*, and includes all the subsequent shows in this multiverse, including *The Flash* and subsequently, *Supergirl* and *Legends of Tomorrow*, *Batwoman* and *Superman & Lois*. Although *The Flash* and *The Arrow* both exist on the same

Earth (just in different cities, or we might say on the same timeline), dubbed Earth-1, many of the other superheroes existed in alternate versions of Earth.

For example, the world of *Supergirl* originally existed in Earth 38. *Supergirl* was about Kara Danvers, who, like her cousin, Clark Kent (aka Superman), was sent from the planet Krypton while a child to Earth and had superpowers as well. On some Earths, Superman didn't exist; in others, Supergirl didn't exist. On one Earth, Superman had been killed by none other than Batman, Bruce Wayne.

The various worlds interacted because it was possible, using either the technology built at Star Labs in *The Flash* or specific superpowers that heroes developed, to communicate with, and sometimes even hop between, these different Earths.

Not only does this is set up an almost infinitely expandable narrative structure, with the ability to retcon almost anything, it also set up a unique set of crossover events, like *Crisis on Infinite Earths*. These crossover events started on a Monday on one show (say, *The Arrow*) by involving timeline or Earth-jumping, and continued on the next show (say, *The Flash*) on Tuesday night, and so on throughout the week. These stories got very complex and it could be hard for anyone older than a ten-year-old or a teenager to keep track of who was from which Earth.

In *Crisis of Infinite Earths*, there is a unique individual, the Monitor, who can see what is happening on different Earths and on the timeline overall. Unlike previous challenges, when a particular Earth is in danger, in *Crisis*, the existence of the entire multiverse is in danger from someone named the Anti-Monitor, who goes back to the beginning of time and finds a way to destroy not just one earth, but all the earths in the multiverse before they branch out as separate timelines.

To try to save us from this rather unglamorous outcome, not only do we get the various heroes from the different shows that we've mentioned (The Flash, Supergirl, Superman, etc.), but we also see *different versions* of the same hero: one Superman, for example, is also played by Brandon Routh, who starred in the film *Superman Returns; another* by Tom Welling, who starred as a young Clark Kent in the TV show *Smallville*, setting up an epic overlapping of stories and personalities and actors.

Spoiler alert: After the earths are destroyed, some of the heroes are able to transfer out of space and time to the vanishing point, from which point they can return to the dawn of time to try to reset the timeline. This results in timelines running forward again, but some things are very different on the new Earths that emerge. Some of these are small changes and some are big changes. For one thing, although most of the earths are still around, several of them, namely Earth 38 (where Supergirl lived) and Earth 1 (where the Flash, the Arrow, and the Legends of Tomorrow were), are all merged into a new timeline called Earth-Prime. Not only are these superheroes now on the same Earth (and no longer require multiverse hopping technology to see each other or to work together) but there's a whole new history of cooperation between superheroes who were previously on different Earths!

This is not only a good fictional example of what a multiverse might look like, complete with different versions of our planet and different versions of ourselves, it also demonstrates how the same timeline run again might result in a very different outcome, resulting in Mandela effect-like situations where different people remember a different history. For example, on the new Earth-Prime, Lex Luthor is no longer a villain but an ally of the superheroes ... well, sort of.

Chapter 7

The Nature of the Past, Present, and Future

What, then, is time? If no one asks me, I know; if I wish to explain to him who asks, I know not.[88]

– Saint Augustine

In a sense, having multiple parallel worlds that are branching off based on quantum measurements implies that there are multiple futures, future versions of us, each of which follows a different path in the graph of possible futures.

But what does it mean for something to be in the future versus the past? This seems obvious, as Saint Augustine said, as long as you don't try to be specific and try to explain it. It turns out that many of the laws of physics are time-reversible: they work equally well whether you are moving forward in time or backward in time. The only thing that prevents us from moving backward is the second law of thermodynamics, which says that entropy (or disorder) increases over time.

But could we also have multiple pasts? In this chapter, we delve deeply into the definition of time and space in physics in general, but most importantly into quantum definitions of time. In quantum mechanics, we started with the particle-wave duality, which led to the counterintuitive idea that a particle isn't in a location but is spread out in a probability cloud or wave. Each of those possibilities, some scientists have concluded, could be in its own separate parallel universe, effectively a separate timeline moving into the future. In this chapter, we'll ask whether the past is also spread out in some kind of superposition or probability wave as well. If so, this gives us a strange view of how time works.

What Relativity Says About Time

In Einstein's special relativity, time flows differently, depending on the speed at which you are going. This is the cause of the famous time-dilation effect that has made its way into certain science fiction (and is ignored in a lot of other science fiction). Time dilation means that time flows more slowly when you are traveling very fast. If you were to travel at 90% of the speed of light for a day (or a week or a year), time would pass more slowly for you. This means that hundreds of days (or weeks or years) might pass for the rest of the world.

This is best illustrated in the twin paradox. Consider two twins, let's say Alice (the adventurous one) and Muffet (who prefers to sit on her tuffet), who were born at approximately the same time. If one twin (Alice) leaves Earth and goes somewhere at a very high speed (the closer to the speed of light, the more pronounced the effect), then Alice will end up being biologically younger than her twin. If Alice traveled for a year, she'll find she's only aged one year.

But her twin on earth might be many years older. In an extreme case, if Alice were to get very close to the speed of light, many hundreds (or thousands of years) might have passed on Earth, and everyone the twin left behind (including her twin) would not just be biologically older, they'd be dead.

Space-Time Diagrams and World Lines

Einsteinian or relativistic views of time and space can be expressed in a series of space-time diagrams. In science, when we think of space, we naturally think of coordinate systems. The three-dimensional physical world can be thought of as a series of x,y,z

coordinates. When Einstein showed that time and space were intimately related, he did it via equations and his intuition. However, it wasn't that easy to imagine. Einstein's old physics mentor, Hermann Minkowski, was the first to develop the most famous space-time diagrams to illustrate what relativity was saying about particles in motion and was the man most responsible for inking time as the fourth dimension (with apologies to Mr. Hinton, whose concept of a fourth dimension included the directions of ana and kata, as discussed in Chapter 4: A Variety of Multiverses.

A very simple example of the layout is shown in Figure 12. The horizontal axis represents position in space (in this case using just a single dimension, the x axis). The vertical axis represents time. The lines that are not the two axes represent the worldline of an object (or a particle)—that is, how the particle moves through space and time.

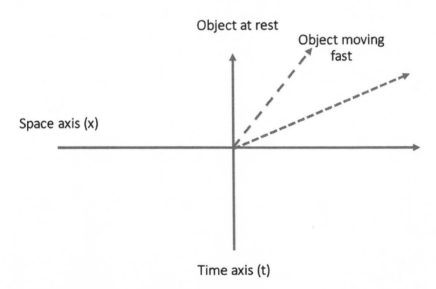

Figure 12: Simple space-time diagram with moving objects.

In a basic space-time diagram, if an object were at rest in space (not moving to the left or right, using our one dimension of space), its worldline would go straight upward vertically (no change in x) while time moves up. If the object were moving, its worldline would drift to the right or left (changes in x) while time marches upward. The faster an object goes, the more the worldline would slope to the right (or left). In this layout, a very fast object covers more distance (x axis) in the same amount of time (t axis).

In these diagrams, time always moved forward, and depending on how fast a particle was traveling, its perception of time (or in Einsteinian terms, its inertial reference point) would change. A particle's worldline was a time-like line that showed its progression through time (moving up) and through space (moving left or right).

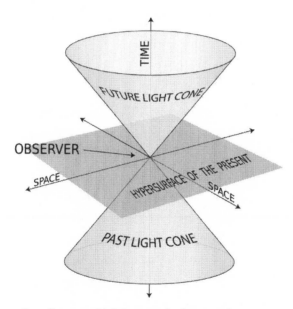

Figure 13: Space-time diagram with light cones for future and past events reachable by light.[89]

Of course, the space part of space-time is vastly simplified in these diagrams as a single dimension. Expanding the spatial dimension from one to two axes (so there is an x and a y axis, or a plane for space coordinates, and still a time axis that is perpendicular to the space plane) results in a diagram like that shown in Figure 13. In this expanded space-time diagram, a light cone is defined by a flash of light that begins at the origin, t=0, and then travels outward across all space at the speed of light. You'll see in the diagram then that the speed of light is defined almost as a diagonal, and the area it encompasses takes on two cone-like shapes, one above the origin and one below the axis. The future areas that can be reached by the light (or by traveling less than the speed of light) are contained within the future light cone, whereas the past light cone encompasses the places from which light could have reached the present moment from the past.

The Block Universe: Snapshots of Space-Time

Since we aren't delving too deeply into relativity in this book, the important point about space-time diagrams for us is that they show us a way of spatializing time and encompassing space in the same diagram. Time is in the picture using coordinates, and this technique is quite useful in getting insight into how the universe might be working and how things might be changing from one moment to the next.

But this isn't the only way to show time in some kind of spatialized way—we'll explore several other ways in Part III (using cellular automata) and in Part IV (using Multiverse Graphs). One approach that is fruitful to cover here extends the Minkowski idea by showing a series of snapshots in time, stacked together. This would be much akin to the view of a movie in which we have 2D frames representing some 3D world next to each other. Each frame comes after or before another frame, and by visualizing them one after the other, we can trace the motion of an object across the screen, as shown in Figure 14. In fact, it was this insight that led Thomas Edison to invent the modern art of movies: showing motion by running frames one after the other.

If we think of time as having discrete values, we can think of each frame of the universe as having one (or all) particles in a particular position, not unlike the pixels of a frame in a movie. The next moment would have the pixels changed slightly and so on for each frame of the movie. This collection of snapshots of space in time would give us the illusion of moving in time, or creating patterns in time and space, even if there is no movement in the objects themselves, only the movement from one frame to the next.

Figure 14: Frames one after another give the illusion of movement [90]

The collection of frames is sometimes called the block universe theory of space-time by physicists, as shown in Figure 15. Generally speaking, assuming that the differences between snapshots are consistent with the (non-quantum) laws of physics, this model fits within Einstein's theories of relativity.

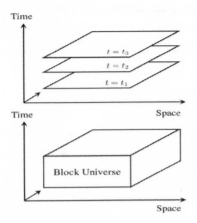

Figure 15: Example of block universe: stacks of snapshots.[91]

When we look at quantum phenomena, we realize that this simple deterministic view of time as it marches forward may be too simple; although we will come back to representing moments in time as snapshots, they may not fit the simple block universe model.

Wheeler and the Delayed-Choice, Double-Slit Experiment

If the idea of time passing in snapshots, let alone particles becoming waves and vice versa, weren't confusing enough, we now come to another confounding aspect of quantum physics that may cause us to redefine our idea of the past and the future.

Just as the wave-particle duality was revealed by the double-slit experiment, our notion of simple time that moves from the past to the future was thrown into doubt by a version of the experiment proposed by ubiquitous John Wheeler. Wheeler wrote an article proposing what he called the delayed-choice, double-slit experiment in 1978.

There are now several accepted versions of this thought experiment, and many more are being created all the time. In the basic version, shown in Figure 16, a laser goes through a slit with a lens, which causes the particles to split one way or another. (The lens is serving the same purpose as the two slits in the original double-slit experiment.) Beyond that, similar to the basic double-slit experiment, is a detection screen. When the screen is present, we see an interference pattern just like in the normal double-slit experiment, with light beam acting as a wave.

However, two detectors are set up to measure whether a photon was sent upwards or downwards by the lens, which would be the equivalent of going through one slit or the other in the original experiment. A photon cannot go both up and down, which means it has to be in particle mode and not wave mode.

Wheeler's double-slit device with delayed choice

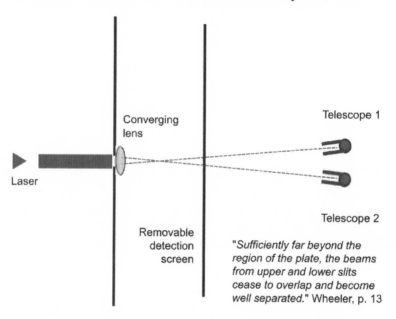

Figure 16: An example of Wheeler's delayed-choice, double-slit experiment.[92]

Obviously, these two receivers cannot measure any photons when the screen is present. However, if the screen is far enough away from the lens, there is a brief moment when the light has entered the lens but has not yet reached the screen. Wheeler proposed that the decision about whether the screen should stay there or be removed not be made until *after* the light has gone through the lens and before it has reached the screen. If the light was going to choose only one direction (and make it to detector 1 or 2), it would have had to make this decision when it hits the lens, which would make it *before* the decision was made to remove the screen.

Wheeler proposed that if the interference pattern shows up on the screen when it's not removed, and the telescopes show the light reaching only one of the two catchers when the screen is removed, this would show something strange. It would show that the wave-particle duality exists even if the decision to measure (or not measure) is made *after* the light has gone through the slit and the lens.

By analogy, this means that a decision in the future (whether to remove the screen or not), is affecting a decision in the past (i.e., which direction to go when passing through the lens).

Wheeler posited this delayed-choice experiment theoretically, but over time scientists have found ways to implement it. The conclusion matched Wheeler's prediction: light retained its wave-particle duality—even if the decision of whether to remove the screen was made after it passed through the lens.

Wheeler concluded that quantum particles are undefined until the moment they are measured, even if the measurement happens *after* the particle makes a choice of which direction to go.

Measurement in the Future (or the Past?)

The strangeness of the delayed-choice experiment, which might also be called a delayed-measurement experiment, was further demonstrated by making the particles in the second part of the experiment travel over a much larger distance before they were measured. This would mean the decision about which direction (or slit in the original double-slit experiment) was made further in the past.

In 2017, a team of Italian scientists did a delayed-choice

experiment over a longer distance by bouncing lasers off of satellites and found that the results were consistent with the smaller-scale experiments.[93]

In this case, a photon traveled thousands of kilometers to a satellite after deciding whether it would act as a wave or a particle (going through one of several slits). The measurement in such a case is clearly being done well into the future from the perspective of the time the photon is first sent into the experiment.

However, until the measurement occurs, the researchers found that the photon still exhibited properties of both wave and particle. This means that something in the future (the observation) was influencing something in the past (a choice of whether to behave like a particle or wave when going through the slits)!

Wheeler proposed a cosmic version of this experiment that would make the results even more startling than the 1000 km version that sent the light to a satellite. It's sometimes referred to as the cosmic delayed-choice experiment or the galactic delayed-choice experiment, and a version of it is shown in Figure 17.

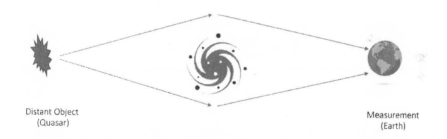

Distant Object
(Quasar)

Measurement
(Earth)

Figure 17: The cosmic delayed-choice experiment.[94]

In this version, a bright very distant object like a quasar is sending light toward the earth. Not only are quasars extremely bright, but they are some of the most distant objects known to us and can be billions of light years away.

Let's suppose there is a gravitationally large object, such as a galaxy or a black hole, in the path between the quasar and the earth, but it's still a million light years away from us. This would naturally bend the light, and the light would have to decide about whether to go around it in one direction or the other. (In reality, this would be a three-dimensional object, so there are more than two options, but let's say the light is polarized either way.)

Suppose we had set up two telescopes here on Earth and pointed them at the galaxy so that one telescope can catch only the light that went left, and one can only catch the light that went right. Once again, the decision of whether to go left or right should have been made in the distant past—perhaps a million years in the past! The decision of whether to measure it is made here on Earth by setting up the telescopes a million years in the future (from the point of view of the light from the quasar when it makes the choice to go around the galaxy to the left or right).

If this doesn't surprise you, then to paraphrase Niels Bohr, you may not have understood it fully. A particle would have had to make its decision *a million years ago*, but quantum mechanics is telling us that decision is made now, in the present, upon measurement.

The Past and the Present and the Future

In the classic materialist model of the universe, the past has already happened and it cannot be influenced by anything that happens after it. Yet (as if quantum mechanics wasn't weird enough already) the delayed-choice experiment seems to be the death knell of this materialist model and raises big questions. Does the past exist? Can it be influenced by the present (which is, in fact, the future as far as the past is concerned)? If so, is the present being influenced by the future?

Some physicists have dubbed this concept *retrocausality*, which means the cause is in the future and the effect is in the past, though it remains an area of debate. Wheeler himself didn't agree with retrocausality, though he perhaps took on an even stranger idea: the past does not exist until it is observed.

Wheeler concluded that: "no phenomenon is a phenomenon until it is an observed phenomenon ..." and that "the past has no existence except as recorded in the present," and that the universe does not "exist, out there independent of all acts of observation."[95]

In some ways, this is butting up against the idea of Schrödinger's multiple histories, which all exist simultaneously until a measurement is made.

These are big ideas that get at the questions about multiple timelines that we are exploring in this book. If multiple timelines exist, that would mean there are not just multiple futures (which most people might agree is possible) but also multiple presents (which common sense dictates can't be the case) and, finally, multiple pasts (which seems completely nonsensical)!

Snapshots from the Multiverse

Perhaps we can refer to our attempts to spatialize time to understand the idea of multiple presents and multiple pasts better to incorporate this idea.

Let's return to the block universe collection of snapshots that we explored earlier in the chapter. How would such a collection of snapshots need to be changed to deal with the quantum phenomenon and the quantum multiverse?

Oxford's David Deutsch, in exploring this idea in his book, *The Fabric of Reality*, argues that the snapshots in a block universe generally would be consistent with classical mechanics, but once we introduce quantum mechanics, things become much more complicated. This is because the outcome of quantum measurements is truly random (i.e., you can't predict which way a particle will go in advance via formula like Newton's laws). So, for each choice at time t, there are two possible snapshots of the next snapshot at point $t+1$. Since both of these future snapshots *exist* because of superposition, a single view of each point in time isn't technically correct in any diagram.

Deutsch argues that we could think of a multitude of snapshots at each time *t* rather than of a single snapshot. These same-time snapshots might be placed next to each other (let's say on the same horizontal plane). The next snapshot (at time *t+1*) would be placed vertically above a snapshot. Looking just up and down at a single snapshot, it would look like the classic block universe: a single timeline of snapshots.

You could theoretically then have a multiverse defined by snapshots. By taking all the snapshots at a particular horizontal plan (or time *t*), you can see the state of all the possible multiverses at that time. By looking up, you could see the possible next snapshots, and by looking down, you could see the possible previous snapshots. Taken together, these would constitute super-snapshots of all the possible universes.

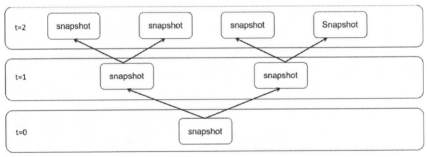

Figure 18: Example of snapshots above each other at time=t.[96]

Deutsch writes: "Physical reality at a particular moment would be, in effect, a 'super snapshot' consisting of snapshots of many different versions of the whole of space." A simplified version of his idea of snapshots in time is shown in Figure 18.

Moreover, these snapshots, writes Deutsch, wouldn't let you predict what will happen, but what might happen. Deutsch concludes that this isn't a "quantum theory of time" as much as it is a way to visualize a multiverse using quantum ideas and concepts.

If we think of the world as a series of snapshots, but not

necessarily in the linear order of a block universe but in a quantum sense, what would these snapshots look like? In fact, within video games, we refer to these snapshots of the world as a gamestate. We'll look at how we build digital snapshots of the world in Part III, including a look at quantum computing and how it figures into the idea of a simulated universe (and multiverse). Then in Part IV, we'll look at how to arrange these snapshots into a complex graph that allows for multiple paths of navigation, resulting in a different way to think about timelines.

For now, we must realize that quantum mechanics in the form of the double-slit experiment, the quantum multiverse theory, and the delayed-choice experiment are all telling us something very strange about the nature of our universe and the nature of the past, present and future.

STAR TREK: THE NEXT GENERATION AND YESTERDAY'S ENTERPRISE

In the "Yesterday's Enterprise" episode in *Star Trek: The Next Generation*, Captain Picard and his crew on the *Enterprise-D* come across an anomaly (the kind of term that's often used but rarely explained in science fiction). Somehow, through this anomaly, they encounter a time-traveling version of its predecessor, the *Enterprise-C*. In the existing timeline, this previous *Enterprise* had been destroyed by the Romulans while responding to a distress call from a Klingon outpost.

In what we might call a timeline switch, the *Enterprise* finds itself suddenly in a new timeline, where the Federation is at war with the Klingon Empire. Of course, none of the crew *know* they are on an *alternate* timeline. They have what we might call timeline amnesia, though to be honest, that's not quite accurate. Technically, the events of the previous timeline didn't happen in this particular instance of these individuals, though somehow we think of them as part of the same individuals. The Captain Picard we see in this new timeline is a wartime captain of a warship.

So how does the *Enterprise* crew know that something is amiss? They don't. Except one person. Only the alien Guinan (played by Whoopi Goldberg), who has a kind of sixth sense about reality and timelines, sees that something is wrong, and that reality has changed. In particular, she finds Tasha Yar, a character from the first season of *ST: TNG* who died and has a visceral reaction. She doesn't know Tasha, but the Tasha instance we see in this timeline knows her. Guinan meets with Picard to tell him that "something is wrong" and things aren't as they are supposed to be. In one scene, she tells Picard that there should be children on the ship, which of course there aren't in this warlike timeline. When he asks her to be more specific, she can't, but she has this vague sense of the way things should have been in the original timeline.

Coupled with this realization that something is wrong, Picard wrestles with the decision of what to do with the *Enterprise-C*: whether to send it back in time through the anomaly, where it would most certainly be destroyed. It turns out that this decision is what it all hinges on. If the *Enterprise-C* goes back and gets destroyed defending a Klingon outpost, the war with the Klingons would never have happened. It also turns out that the war isn't going well and the Federation is on the verge of being defeated.

They make the decision to send the *Enterprise-C* back in time, along with Tasha Yar, who didn't belong in the future. The *Enterprise-C* is destroyed, as originally happened, by the Romulans, and the Federation never starts the destructive war with the Klingons.

In short, we find the *Enterprise-D* to be back on its original timeline. The Tasha Yar that was sent to the past would now be classified as a time remnant, a remnant of a timeline that no longer exists.

In this timeline, in the present timeline of the *Enterprise* in *Star Trek: The Next Generation*, only Guinan retains vague memories of the other timelines. As Philip K. Dick said, "[S]uch an impression is a clue that in some past time-point, a variable was changed—reprogrammed as it were—and that because of this, an alternative world branched off."

In a strange consequence of this time travel, and to confuse things more, in a later episode, we learn that Tasha Yar was sent to the past and had a child with a Romulan father. For Picard and the rest of the crew, this seems absurd, since in our timeline, Tasha Yar died earlier. But because of the time remnant from the other timeline, how she was sent back to the past, her daughter is now all grown up (and consequently played by the same actress, Denise Crosby).

Part III

Building Digital Worlds

The history of the universe is, in effect, a huge and ongoing quantum computation. The universe is a quantum computer.[97]

—Seth Lloyd, *Programming the Universe*

Chapter 8

Multiple Timelines in *SimWorld*

Thus far, the physicists' models for multiple universes either deal with multiple *actual physical* universes or a wave of probabilities that collapses into a single physical world. Information theory was nascent when quantum physics first became a thing, but over time more and more physicists have started to realize that the information carried by particles may be just as important as the physical particles themselves. In fact, the more that physics delves deeply into subatomic particles, the less they can find an actual thing called matter, leading to Wheeler's famous phrase, "it from bit," which we spoke about in Chapter 5.

As I stated in earlier chapters, both of the most popular interpretations of quantum mechanics (the MWI and the Copenhagen interpretation) are more likely in a digital universe than in a physical universe. By *more likely*, I mean we can rely less on abstract concepts like magic (as in, the universe just magically splits off into multiple universes, or the wave function magically collapses), but rather deal with processes that we are at least capable of understanding and modeling.

In this chapter, we'll use simplified computer gaming techniques, and in the next chapter, we'll look at simplified computer programs called cellular automata to explore how simple digital multiverses might be implemented. Don't be fooled by the fact that these techniques are very simple and don't require much knowledge of programming or computer science. They are still very powerful.

Finally, we'll end this part of the book with a chapter on the exciting potential of quantum computing, which taps into the very fabric of the universe (and the multiverse) to accomplish incredible computational tasks. This can give us insights into how a simulated multiverse might work.

Laying the Foundation for a Digital World

Let's back up to the very simple world of computer games, circa the 1970s and early 1980s. Although arcade games were becoming popular, it was text adventure games that represented an important step on the road to the simulation point. Text adventures were the first games that really gave us a sense of a large and complex world that could be explored inside the computer.

Games like *Colossal Cave Adventure*, the first text role-playing game (RPG), inspired a whole genre of wildly popular text adventure games like *Zork* and *Planetfall*. More important, they inspired a later generation of graphical RPGs that began to use pixels in a rudimentary way to represent this rich virtual world that the player could explore.

I have a particular affinity for these types of text adventure games, even though I haven't played one in ages. This is because the first games I ever programmed beyond Tic-tac-toe were of this genre, based on a book I managed to get my hands on while in junior high school about creating text adventures in BASIC.

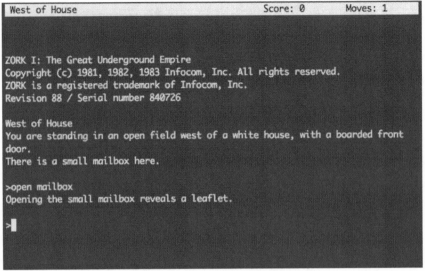

Figure 19: Screenshot of Zork: The Great Underground Empire.

In my earlier book, *The Simulation Hypothesis*, I identified several key elements that laid the foundation of a game world that could be explored in text adventures:[98]

- A Big Game World (represented by a World Description)
- Player Game State
- World Game State
- Non-Playable Characters
- Player Commands

Although building games isn't the subject of this book, showing how we represent even a very simple game world will help us understand how a digital universe and, ultimately, a digital multiverse might work.

Inside the Game World

These textual game worlds, which often resembled campaigns of the *Dungeons and Dragons* tabletop game, consisted of a set of discrete places or rooms in the game world that the player could navigate. In the original text adventure game, *Colossal Cave Adventure*, these rooms were caves of the underground cave complex (based partly on Mammoth Cave in Kentucky, which the original author, Will Crowther, had enjoyed exploring).[99]

In these games, each room had a text description and could contain objects. The player could navigate using text commands or pick up these objects (which put them into the player's inventory) and then drop them in various locations in the world, thereby affecting the states of both the player and the game world.

These commands (such as *examine*, *pick up*, *drop*) were typed in via the keyboard, along with navigation commands (such as *go north*, *go south*, *go east*) that took the player from one room to another. The game designer usually had a map that connected the different rooms, but the player was forced to redraw this map by exploring. Figure 20 shows a hand-drawn version of the map for *Colossal Cave Adventure* that eventually made its way onto the Internet and gives you a pretty good idea of the map that could be explored.

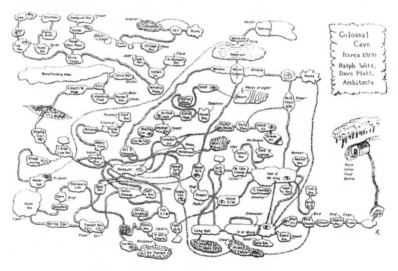

Figure 20: A hand-drawn map of Colossal Cave Adventure.

What we're most concerned about here is how the *gamestate* is represented and how it changes when the player takes actions. Why? Because we want to understand how one world can be represented digitally; we can then understand how to move from state to state, thus defining timelines in a digital world.

Representing the World

We will create a very simple text adventure game skeleton, called *SimWorld*, and explore how we might come up with simple representations of the gamestate.

For *SimWorld*, let's pretend we have 15 rooms that we could go to, with a sample map as shown in Figure 21.

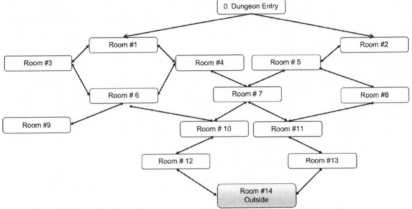

Figure 21: The game world of SimWorld, a simple map of rooms.

Note that the player starts in a dungeon (room #0) and can explore the dungeon by going in a cardinal direction such as N(orth) or S(outh) or E(ast) or W(est) or, in some cases, U(p) or D(own). These last two (Up and Down) represent moving through trap doors and openings in the caves that surround the dungeon. Let's suppose the goal of *SimWorld* is to end outside of the dungeon, which we have defined as the 15th room (or room #14, since the first room was #0).

What data do we need to keep track of to implement such a game? The three main elements follow. The first one is *static*, whereas the other two are dynamic: they change as the player issues commands and the game goes on. Collectively, we could consider these the full *gamestate*.

- **World description** (*static*). This defines the virtual world the player can explore, a digital version of Figure 21. Most notably, it contains a description of every room. In the earliest games, this was stored as text on disk, whereas the first (slightly more sophisticated) graphical RPGs had one picture per room. Later, graphical MMORPGs used 3D models to represent the world, and it was rendered as needed based on the player's location, allowing for continuous movement around the world. The world description also contains information on the links between the rooms (i.e., the map), usually stored as some type of array/table structure, or what mathematicians might call a graph of nodes.

- **Player state** (*dynamic*). What information about the player do we need to store in a text adventure game? The obvious thing is which room the player is in. Also, we might want to have a way to measure the player's performance, perhaps a level that goes up at certain milestones. In our simplified game, we'll ignore level. We will also allow the player to pick up objects and put them in inventory.

- **World state** (*dynamic*). The only thing we really need to keep track of in our initial world state is the location of objects that are not in the player's inventory. Picking up and dropping objects are a very simple example of the player changing the world. You can envision many more complicated ways that the world could be changed by the player, including opening and closing doors, etc.

Objects, which are perhaps the only complicated aspect of *SimWorld*, can be located in any room or in the player's inventory, which means they would be either part of the player state or the world state.

Representing the Gamestate

The two main tasks in building a simple video game involve defining the data structures and writing the code. Although writing code would vary drastically, depending on which programming language, operating system, or hardware the game was being built for, the data structures pretty much remain the same across games, which is why we will concentrate on that.

SimWorld has fifteen rooms, each with its own name and text description. Let's also decide that *SimWorld* has only four objects: a hat, a bat, a cauldron, and a spellbook. Since the map doesn't really change, it isn't quite as interesting to us as the dynamic part of the gamestate, which consists of two parts:

- Player location
- Object location

Although today we would use a higher-level language like C# or Python or JavaScript to build these games, and data structures like JSON or NoSQL or SQL, the languages and data all eventually come down to bits that are either stored or processed.

Usually, the task is about optimizing so that you can represent the information you need with the least number of bits. This was particularly true in the old days of computing when memory was precious and storage was limited to floppy disks or cartridges. But it's also true in today's Internet-connected, multiplayer games. Whereas storage and processing are plentiful today, the costly operation is now sending bits over the wire (for example, over a wireless data stream from your phone). A *bit* (short for "a binary digit") is considered the simplest form of information and can have a value of either zero or one.

Let's look at how we might represent this dynamic gamestate in a digital world. Experienced programmers might laugh at how simplified this format is, but the reason we are focusing on bits will become clear as we move from a single digital world to a digital multiverse.

Player location: Since there are 15 rooms, we would need 4 bits to represent which room the player is in. In computerese, we refer to 8 bits as a byte and to 4 bits as a half-byte. Consequently, a half-byte could represent rooms #0 (in binary with 4 bits: 0000b) to room #14 (or 1110b)[100]. The b at the end of a number tells us that it's a binary representation of the number. Programmers will see that using 4 bits is natural, but there is an extra value, the 16th value, or 1111, which we aren't using (yet).

Object location: If there are 4 objects total, you could represent a full list of them with 2 bits: object #0 (or as two bits, 00b) to object #3 (or 11b). However, for each object, we have to keep track of its location, or which room it is in. This requires 4 bits (the same as the player location), but we have for objects, so this requires an array (or list) of 4 locations, one for each object, like this:

- Object(0): hat is in room #0 (or 0000b)
- Object(1): bat is in room #2 (or 0010b)
- Object(2): cauldron is in room #14 (or 1110b)

We usually reserve a value to designate a special value. It turns out we can use our extra room, the 16th room, (or room #15 if we start with room #0) to represent when an object is not in any room in the map but is in the player's inventory instead, like this:

- Object(3): spellbook is in room #15 (or 1111b)

There's nothing special about using the number 15 (1111b); we could just as easily have started the room numbering with 1 rather than 0 and used 0 as the special inventory room.

The World in a Word

It used to be that you could only play a video game during one session. If you ended the game, you had to start over from the beginning. A big innovation came when you could save your gamestate and restart playing from the place where you left off, rather than starting all over again.

To do this, we need to do what's called serializing an object to save it somewhere (either on disk or on a server). This essentially means saving the bits that are in memory in a format that can be reread and reinterpreted later. To do this, we have to agree which bits go where and what they mean (i.e., the first 4 bits represent the player location, the next 4 bits represent the location object #0, the hat, etc.).

Anyone familiar with video games will realize that so far we have an almost childishly simple gamestate; obviously, this isn't enough to make much of a game, but we are simply illustrating the concepts of a gamestate, not making a real game.

To represent the gamestate we have just described, we could use several bytes of information, with each chunk being stored in a half byte (or 4 bits).

We could now represent the whole gamestate, which could be saved in memory or on disk, with only 3 bytes, or 24 bits total. We could also easily send such a small number of bits over the Internet to a server, where they could be downloaded to a different computer, which, if it had the code for *SimWorld*, could continue to play the game from the same point forward without skipping a beat.

A digital representation of the gamestate that we just discussed is shown in Figure 22. It shows the player in room 5 (in binary, 5 would be 0101b), with the hat and cauldron in the player's inventory. The whole thing could be represented in a string of bits like this: 0101 0000 1111 0100 1111 0011b. It is broken out graphically so you can see how it's all arranged by bytes (and by half-bytes) in the table in Figure 22.

Byte #1		Byte #2		Byte #3	
1st Half 4 bits	2nd half 4 bits	1st Half 4 bits	2nd half 4 bits	1st Half 4 bits	2nd half 4 bits
(player location)	(unused)	Object(0) Location (hat)	Object(1) location (bat)	Object(2) location (cauldron)	Object(3) location (spellbook)
0101	0000	1111	0100	1111	0011

Figure 22: Simple gamestate by byte.

We now have a very compact way to represent the entire dynamic game. In fact, we really only needed 20 bits, because we have 4 bits left over (the second half-byte of byte #1), which we could use to store the level of the player or some other information.

In computer-speak, a *word* is usually a collection of a number of bytes (and, by extension, of a number of bits) that defines a basic chunk of information that the computer can load and process fast. The size of a word can be 4 bits, 8 bits, 16 bits, 32 bits, 64 bits, and so on. In our case, three and a half bytes represent our basic gamestate, so we could literally put the whole world into a word of 4 bytes (or 32 bits) and still have bits left over for expansion!

Adding an NPC and Some History: The Ogre and the Doors

Let's suppose we wanted to add an NPC, or nonplayable character, such as an Ogre, to the game. Unlike Ogres that want to attack you, let's suppose that he's just a grumpy Ogre like Shrek, who doesn't really like you and just wants you to leave his dungeon. As a result, he's willing to facilitate your departure by opening certain locked gates and doors.

Let's also suppose the ogre can wander around to different rooms just like the player can, and if you are both in the same room, the Ogre will unlock one of four doors for you, just to get rid of you. The location of the four doors would be stored in the world description, but the state of each door, as well as the location of the Ogre, would both need to be in the gamestate itself.

In a text adventure game, you might see text like the following if you end up in the same room as the Ogre.

```
>You are in the lower dungeon. To the west
is a passageway and to the east is a door
that is closed and appears to be locked.
>You see an Ogre, who is standing in the
dark looking at you, expectantly pointing
at a (locked) door to the east.
```

Since we aren't getting into the code of the game for this exercise, and all we care about is this representation of the gamestate, how could we modify our gamestate to accomplish this?

We would need to store the Ogre's location, which requires another 4 bits, and a way to keep track of the state of each of the doors. Since there are 4 doors and they only have two states (locked or unlocked), this can be easily represented with 4 bits, 1 bit per door. A value of 0 for a bit means that door is locked (default value unless the Ogre helps you), and 1 means that door is unlocked.

We now have one byte added to the gamestate, and we have our word for the whole simulated universe contained in 32 bits:

> Byte #4 – 4 bits for the Ogre location (let's say 0101) and 4 bits for the doors (which start out at 0000 when they are all locked and 1111 when they are all open).

Saving and Restoring the Game

When save game and load game were first invented, the idea was just to have one saved gamestate per player, so that if you were halfway through the game and you needed to turn off the computer or go eat dinner, it would still be there next time. One complicating factor was that these games were originally run on mainframes, so there might be different players at different points in the game but playing on the same machine, so gamestates had to be stored in files that were distinct.

In the very first graphical computer games, such as *Space War* (invented in 1961) and *Pong* for Atari in 1973, there was no save gamestate. Saving a graphical gamestate would by nature require more bits than a textual adventure game, but as memory and disk power grew, there was no reason this couldn't be done as well. For example, in *Space Invaders*, shown in Figure 23, you could save the gamestate by having a state of each of the aliens, of which there are 11 in each row and 5 rows total.[101]

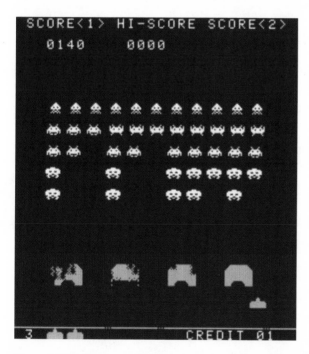

Figure 23 The classic arcade game: Space Invaders, *5 rows of 11 aliens each*

Now *SimWorld* is so simple that you may not care to save the gamestate, but if we add more NPCs (Dragons and Elves maybe?), more rooms (say 256 or 1024 or, as in the game *No Man's Sky*, 18 quintillion planets), and more obstacles (not just doors but other puzzles to solve), then it might be difficult to finish the game in one session. Even for *SimWorld*, we might store different gamestates to represent different points of progress in the game:

- V1—level 5, in lower dungeons, Ogre has opened 2 doors
- V2—level 6, in lower dungeons, have gotten key from the dragon
- Etc.

Saving multiple gamestates is actually a key part of the reason I went through in detail how we would build a gamestate out of bits. It is also how a multiplayer game would store multiple gamestates on the server, one for each player.

Context Switching

So, what does this representation of a gamestate have to do with our discussion of a multiverse and multiple timelines? The reason I went through so much detail in how to represent a gamestate in a compact way is so that we can look at how those bits might be used.

If we were in a graphical adventure game, the items would be rendered graphically in each room only when you, the player, or rather the player character (PC), goes into those rooms. And the Ogre or other NPCs will only be rendered when they are in the same room as you; otherwise, the objects and the NPCs exist only as information that is waiting to be rendered. The gamestate is the information that is rendered for you, the player.

The point is that you could be in the middle of a game and then suddenly load a previously saved gamestate. Not only would the state of the world change (object locations), but the player's character and all the NPCs would suddenly be inserted back to a previous point in the game.

This process, of saving and loading a gamestate, is analogous to a process that we call *context switching* in computer science. This actually becomes a way to have multiple timelines within even a very simple game like *SimWorld* and switch between them.

Although I had learned to make simple text adventure games on my own, using first my Commodore 64 and then later my Apple IIc home computers (with a little black and white TV as my monitor), it wasn't until I got to MIT that I learned about context switching and multiple processes. By that time, multiwindow graphical user interfaces (GUIs) like in the original Macintosh were popular and I was curious how the processor could have multiple windows running at the same time. Everything I had learned about programming at that point was single-threaded, and though I didn't know too much about the internals of microprocessors, I was pretty sure they could only do one instruction at a time.

The mystery was cleared up when I learned about context-switching, which is defined as:[102]

> In computing, a context switch is the process of storing the state of a process or thread, so that it can be restored and resume execution at a later point.

This means that if there are two windows (let's say Microsoft Word and Microsoft Excel) running on your screen, but you only have one microprocessor, or central processing unit (CPU), the CPU can only be running one process (or one context, as it's referred to) at one point in time. It runs that process (or code base) for a few steps, then it stores the context somewhere else (say on disk) and switches to the second context. The second process runs for a few cycles, and then it switches again.

During the time one context is loaded, the processor is unaware of other processes or contexts, which are stored somewhere else. The time slices between context switches are small enough that from an end user's point of view, it seems like both processes are running simultaneously even though there is only one running at a time.

Context Switching in *SimWorld*

So, what does context switching have to do with *SimWorld*?

Well, the point here is that loading a gamestate into *SimWorld* is kind of like context switching of a CPU. Although we have only one set of code, we can have any arbitrary versions of the game context (which we called the gamestate) that can be saved, stored, and reloaded, at which point the game picks up right where it left off.

No one inside the game world would realize that the context has switched, any more than a word processor knows that you took a break to go to the bathroom and browse the web. This is particularly true of NPCs. In fact, the Ogre wouldn't even know that anything different had happened; it would start moving around just as it was doing the last time this game was played. Assuming that the gamestate said it had opened only one door, that's how it would proceed. This could be the case even though we were in the middle of a game where the Ogre opened three doors for us when the context switch happened.

The Ogre and, more importantly, the player character (PC) might now find themselves in the same place as Philip K. Dick thought that he was: "...living in a computer-generated reality and the only clue we have is when some variable has been changed." Just like Dick's missing light switch (courtesy of his adjustment team), some variable has literally been adjusted by changing a very tiny part of the gamestate. In our case, it may just be one bit that's changed! You would have the sense, as Dick described, of reliving the same moment again, if you had the ability to remember a previous play of the game or the simulated world.

If we think of the amount of information it would take to save the gamestate of our entire world, including the location of every single light switch and person, obviously everything becomes much more complex. Nevertheless, the idea is the same: By changing some bits, we create a discontinuity of some type, and most of the characters (or people in the simulation) would be blissfully unaware of this change. Only those somehow aware of previous runs might even have a clue that something has changed (think Guinan in the *Star Trek: The Next Generation* episode, "Yesterday's Enterprise"; she was the only one who realized that something had changed in their timeline).

Storing History in the Gamestate

In this case, by loading a new gamestate for *SimWorld*, we didn't really change history; we changed the *results* of the past. You'll notice that only a very small amount of history is embedded in this gamestate. By having door 3 open, the implication is that somewhere in the past, the Ogre must have been in that room and opened it. Similarly, if the spellbook is in inventory, this means that the player must have been in the room where it was originally, and picked it up. For lack of a better terminology, you might say that the present moment (the current gamestate) *constrains the possible pasts*.

To store the full history of the world would be more complicated (and require many more bits). One way around storing every possible gamestate in the past would be to store every single move you made in the game. Given only 16 rooms, the number of moves is quite limited, so this isn't a big chore, but you could go explore the rooms for any length of time, which means that the information needed to store history keeps growing the longer that history is. In general, this is something we want to avoid in computer science when possible, because performance degrades when files get large and memory is limited; we like to store the least number of bits to accomplish our task.

Another approach is to store the history, using a simplistic method of storing each room visited. To do this, we could use 4 bits per location (just like we did earlier), and we would get a sequence like that shown in Figure 24:

History in binary	History in decimal
0001,0010,0011,0100...	0, 1, 2, 3, 8, ...

Figure 24: An example of storing history by storing every room visited.

You can easily look at both of these ways of storing history (the moves vs. the rooms visited) and realize there are many ways to optimize the number of bits required to store the history. As a simple example, since there might be only four commands, we could use two bits to store the commands rather than the rooms: E, N, S, W. You could also realize that the map of the world limits the possible futures and histories. If from room #12, for example, the player can only go North or South, then there are only two possible moves. This is also a way of constraining the past. There are of course, an almost unlimited number of ways to try to optimize what bits are stored, which is what happens every day when you send a JPEG file or a video file (say an episode of Game of Thrones) over to your tablet. The bits are compressed, and then at their destination, they are decompressed to be rendered on the destination's computer system.

Timelines in *SimWorld*

Perhaps more important than the specific method of compression is that we can store the entire history in the gamestate (in some compressed format). By doing this, we have defined not just a point in time that we can reload, but a complete timeline or a specific instance. A timeline in *SimWorld* then could be defined:

TIMELINE = <GAMESTATE>, <PATH/HISTORY>

This is actually a partial timeline because it refers only to the past, not to the future, since we are storing only rooms visited.

If we were to draw a graph of all the possible histories, we would see that the map and the current gamestate constrain what histories are possible, both by which rooms are accessible from other rooms (the map) and what choices the player has made to get to the current gamestate (the history). This is a very simple analog to what we will call a Multiverse Graph in the next part of this book, because in a way, the rooms/locations in a text adventure represent all of the possible nodes or futures, and the timelines represent how we are traversing this graph. Two timelines would be represented by two paths through these nodes.

The idea that choices constrain the world can be used in a trick mathematical way to understand a past without storing every single step of the way. The way you do that is to encapsulate the previous past into the current gamestate, using a key or code that is modified each time. You can then determine by looking at the code and, given a limited number of possible pasts, you can decide which one matches the current code. We won't get into the mathematics of how this is done, but this is essentially how cryptocurrencies like Bitcoin validate blocks on the blockchain. Each block is encoded with some information from the previous block. This limits the possible pasts of a particular block and lets anyone validate that the current block is valid.

Replaying the Past: A Form of Time Travel?

We now have a way of thinking about timelines within our simple *SimWorld*. Or rather, we have a digital way of thinking about possible timelines and a way to store our gamestate and revisit any past point in the simulated world.

If we were running *SimWorld* as a simulation, we could run it multiple times and figure out which way people might go. We could arbitrarily save the gamestate at a particular point, and we could then rewind the game to that point and rerun it, perhaps changing variables along the way. We have the perfect analog for how a simulated multiverse might work, complete with context switching as a form of time travel, using only classic computing.

If we are at a particular gamestate, note that the possible futures will be clipped based on the choices we have made to date. It is also possible that a particular gamestate has constrained possible pasts, much like the example of the cosmic delayed-choice experiment in Chapter 7, in which the light might have passed on either side of the black hole. We won't know which one is chosen in the current timelines until the light is observed. You might say that in *SimWorld* we don't know unless a specific history is chosen and loaded—either as part of the gamestate or calculated from the present looking backward.

NPCs, PCs, and the Mandela Effect

Each time the timeline is reset, or context is switched, what would happen? As we said earlier, the NPCs would know the history only up to the point of the reset (think of our example of the unlocked doors). If we were residents of *SimWorld*, it's as if we were magically placed at a different point in the timeline and no one would remember the other timelines.

However, there are several exceptions to the expected behavior in which no one remembers the other timeline. On the one hand, if we were PCs, we get into the first of these exceptions: we might remember the other timeline or have a strong sense of déjà vu. As Philip K. Dick said, we would have "the sense that we were reliving the same moment again and again," although in this case we might say that we are watching our avatars relive the same events again and again. In fact, the whole purpose of playing again would be to make different choices than we did last time. We see there is a meaningful reason to rerun a particular part of the game: to find the optimal path forward.

There's another more practical reason we might not store complete histories in every character but reconstruct those histories dynamically, which can result in differences. As I've said repeatedly, designing any computer system is usually about optimizing resources for maximum efficiency. This means not storing duplicate information. If history A exists inside both timelines, B or C, it makes sense not to store it twice but simply to point to A from both B and C.

Similarly, if we are storing multiple versions of the same characters (what I like to call "time instances" of a character X), then it would make sense to try to optimize the storage of those time instances by storing information that is shared only once, rather than duplicating it for each instance. If we duplicate the history, we can have errors or even different histories for each instance. A simple glitch might account for loading the wrong timeline or memory for character X, or it might be a faulty bit that gets loaded.

> **Time instance** – a time instance is a particular version of a character or object that may be different across timelines, but shares some basic identity.

Moreover, if we simply infer history A from loading B and C without needing to reference it explicitly, that is even better. A more interesting possibility is that if each character calculates the possible pasts each character (or rendering device) might choose to resolve the past graph differently. As long as both are possible valid pasts, if the algorithm that resolves the past is similar to quantum mechanics, then each character is choosing the past that seems most optimal to him (or her or it) in the moment.

Multiple Players in *SimWorld*

To make things more complicated, let's suppose we are in an MMORPG with multiple players on multiple machines using the same server. The techniques for this are very similar, whether it's a multiuser text adventure or a 3D graphical game like *World of Warcraft* or a mobile game like *Fortnite*. An example of a multiplayer architecture is shown in Figure 25, where each player renders the world on their own rendering device (a phone, a PC, a text console, a VR headset, etc.).

Each player has a character gamestate while there is theoretically only a single world gamestate that is maintained on the server. Each player's machine loads a gamestate by requesting it for their character. But as part of a multiplayer game, our characters are often in rooms with other characters, so we request the states of these other characters from the server as well, including NPCs and other PCs. We can store different versions of our characters on the servers also, just like we discussed in a more complicated version of *SimWorld*. Since we only had one character in *SimWorld*, we didn't distinguish between player gamestate and character gamestate.

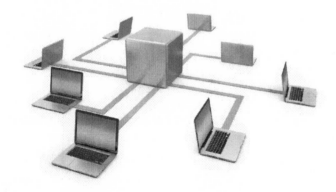

Figure 25: A basic multi-player architecture.[103]

Now we raise the possibility that player x and player y have many versions of their main characters, X and Y, all stored on the server, all captured at different points in the game play. So theoretically, they could be from the same timeline at different times, or from different timelines.

Context switching in this case is simply loading your character's information from the server. The character gamestate and world gamestate are sent from the server down to your local device, which is what happens every time you log in to the game. From the point of view of character X or character Y, they just take whatever context is downloaded from the server, oblivious that other characters may or may not have gotten the same history. All that has to be consistent is that the current state of the world has to be the same; the way the characters got there need not coincide.

You can now easily see a situation where the server sends character X_2 for player x and Y_7 for player y. There is no reason that can't be done, as long as both characters can be loaded into the current state of the world but have different histories. We now see a mechanism where two characters could be in the same scene, but remember history evolving differently – a digital Mandela effect!

What Can We Learn from a Simplified Digital Multiverse?

In this chapter, we explored how we might build a very simple toy universe: *SimWorld*, which uses very simple gaming techniques that have been used going back to the dawn of the current computer age. *SimWorld* was created simply to illustrate how to keep track of a snapshot of the world in digital form and how that snapshot might relate to possible pasts and possible futures.

Moreover, this ability to store and then restore a previous point anywhere in the game brings up the idea that for those inside the game, context may be switched, and they have no idea that they are reliving a specific moment or that some things might have been changed. Like a microprocessor, the game just runs its code using whatever context (data or gamestate) is current. We now can see how multiple timelines could be running, or one timeline could be reset and rerun. It also shows how different characters could remember different histories, yet all be part of the current run of the simulation.

Of course, the adventure game metaphor was only meant to give us a rough idea of how timelines could be represented digitally in a simulated universe. Moreover, an adventure game has both deterministic elements (the NPCs) as well as nondeterministic elements (which come from the choices of the player), which can make things more complicated and lead to unpredictable possible futures.

Thus far, we haven't said anything about how a digital world could be made to look realistic and whether the techniques used for this could be used to simulate multiple timelines in a graphically rendered world. In the next chapter, we turn our attention away from adventure games to the science of complexity and cellular automata, a fascinating subject that bridges computer science and mathematics. Looking at a completely different kind of toy universe might give us a different and more nuanced way of how a simulated multiverse might work.

Chapter 9

Simulation, Automata, and Chaos

It's always seemed like a big mystery how nature, seemingly so effortlessly, manages to produce so much that seems to us so complex. Well, I think we found its secret. It's just sampling of what's out there in the computational universe.[104]

–Stephen Wolfram

In this chapter, we'll explore some computing concepts that are proving to be useful in simulating the physical world: cellular automata, fractals, and chaos/complexity. Although we are moving from simple textual worlds to graphical simulations, we will see that at a basic level they are represented as bits and follow certain rules, just like in *SimWorld*. This will allow us to see how multiple timelines and the Core Loop might work in a graphical world, albeit a very simplified one.

Although I didn't specifically call out these concepts as separate stages on the road to the Simulation Point, they are nevertheless important milestones. They build our understanding of how the world around us may consist of information that is being processed using simple rules. This in turn leads to the conclusion that the complex world around us can be simulated by a computational system.

These concepts also open up different ideas about the nature of computing: distributed computing systems that can help us develop insights into artificial life and artificial intelligence. In this form of computing, rather than thinking of a central processing unit (or CPU) running commands, individual cells are following very simple rules. Although the rules seem simple, what we are really concerned about in this chapter is the concept of *emergence*, in which individual agents following their own basic rules produce something collectively that appears more complex and ordered than the sum of their parts. Both cellular automata and their cousin, fractals, demonstrate emergence in a digital world.

This idea of emergence from distributed processing is one element of how the natural world seems to work, whether we are talking about neurons in the brain, cells inside living organisms, or even a colony of ants working together to build an anthill. In fact, much of the progress on AI over the past decades has been because we were able to model individual neurons, which follow set rules, leading to emergent AI that can recognize faces, generate deep fake images, drive cars, imitate voices and even write text. The techniques in this chapter sit at the intersection of mathematics, biology, and computer science.

Biology as Information

The first type of graphical computational system we will examine is cellular automata, where you have a grid of cells that follow specific rules, resulting in interesting and complex patterns.

One of the big mysteries that comes up in the field of cellular automata and in biology is that the individual units *seem* to be self-organizing. This means that they go from a higher level of chaos to a lower level over time. This seems to violate the second law of thermodynamics, which states that entropy (or disorder) increases in the universe as time goes on. Within biological systems, however, as cells grow it seems that they create order of some type—indeed, this is the basis of living things, including humans.

Throughout the twentieth century, leading up to and including the discovery of DNA by James Watson and Francis Crick, it started to dawn on both biologists and computer scientists that biology was based on the fledgling field of information science.

For example, The Royal Society started an article titled "DNA as Information" with: "The biological significance of DNA lies in the role it plays as a carrier of information, especially across generations of reproducing organisms. ..."[105]

In other words, if we can design organisms (artificially) that can use that information to pass on the same (or even slightly mutated) bits of information to their descendants, then we have simulated one of the most significant aspects of life: reproduction.

Von Neumann and the Birth of Automata

John von Neumann was a Hungarian mathematician who made significant contributions to physics and was one of the grandfathers of the emerging field of computer science. In fact, the classic architecture that we use for running any kind of computer program today—a CPU that accesses local memory—is still referred to today as the von Neumann architecture.

The journey of cellular automation began in the 1940s when von Neumann was looking for a way to describe self-replicating machines. He reasoned that the key to producing artificial life was to have a machine that could duplicate information it contained in another machine, which could then do the same.

Initially, von Neumann was referring to *physical* machines. The idea that he first presented in a lecture in Pasadena, California, in the 1940s was very complicated. Stephen Levy, in his book, *Artificial Life*, describes the basic components that made up von Neumann's theoretical self-replicating machines, which he called kinematics (but which are mostly called von Neumann machines today).

The system consisted of raw materials in a lake, along with four components required for this self-replicating machine labelled: A,

B, C, and D. Component A was like a factory, which scooped up raw materials from the lake and used them in ways that were dictated by some data, which we might call a computer program today. Component B was a duplicator that read and copied information from the first machine to its duplicates, in the same way that DNA is passed down from parents to children. Component C was like a computer and controlled who did what, like a central processing unit. Component D was the actual data, or instructions, which in those days von Neumann envisioned as a very long tape.

Needless to say, this technology would be complicated even for us to build today, let alone for the 1940s. Readers of science fiction might recognize something similar that was presented in the Arthur C. Clarke book, *Rendezvous with Rama*, in which humans find a cylindrical object floating through the solar system that is clearly of alien origin. Although no life forms were found, there was a sea of raw materials and some robots that could use those raw materials to create additional copies of themselves.

NASA commissioned studies on self-replicating machines in the 1970s and 1980s which could be used on long interstellar voyages, no doubt inspired by Clarke and von Neumann.

Today's 3D printers have started to make a dent in von Neumann's vision. They are now being used to assemble objects in space from raw materials. This is seen as a critical way we might be able to conduct manufacturing on foreign worlds, as long as raw materials are available. Theoretically, a 3D printer could print another 3D printer, thus realizing von Neumann's general idea of self-replicating machines.

Although von Neumann's vision was considered impractical in its day, at least two very important ideas came out of his effort, which laid the foundation for a new type of computing. In his original lecture, titled "The General and Logical Theory of Automata," and subsequent papers on this topic, von Neumann gave birth to the idea that an automaton is not just a physical machine, but software that follows some set of rules. The second, perhaps even more important idea, was that information and rules could be used to induce self-replication by passing that information on down the line, well before the discovery of DNA.

This makes von Neumann a pioneer not just in computer science, but in biology as well. Freeman Dyson, the notable physicist, remarked about this much later after DNA had been discovered: "So far as we know, the basic design of every microorganism larger than a virus is precisely as von Neumann said it should be."[106] According to Levy, von Neumann had another crucial insight many decades before chaos theory: "that life was grounded not only in information but also in complexity."[107]

Cellular Automata in Software

Science fiction scenarios aside, kinematics was effectively dead because it was impractical until von Neumann had a chance encounter with an old friend, mathematician Stanislaw Ulam, during the Manhattan Project. Ulam suggested that the concepts von Neumann was proposing would be better realized by leaving behind physical machines. Von Neumann should think instead of a logical *entity* that could be replicated, making the whole thing more practical.

Ulam suggested thinking of a grid, not unlike a checkerboard, and of individual squares as cells, each of which had a state. The state would change with each time step, updating itself based on some rules. Using this approach, von Neumann could focus on the rules and the information. In essence, Ulam told von Neumann to focus on building self-replicating life in the digital world rather than focusing on self-replicating robots.

This was the insight that led to cellular automata as we think of them today. Von Neumann attempted to reproduce his complicated kinematics architecture on an infinite grid, and his first attempt was, although less complicated than the physical machines, still quite complicated. For one thing, it required thousands of cells in the tape (containing information) that would be fed into the other cells.

Despite this complexity, von Neumann made progress on creating cellular automata, a logical, self-replicating "life" form, using this grid and cell approach. The name *cellular automata* (CA) itself would only be coined later by Arthur Burks, who contributed to the first digital computer, the Electronic Numerical Integrator and Computer (ENIAC), and was part of the first cellular automata lab at the University of Michigan decades later.

Although von Neumann's design is significantly more complex than the simple CAs we will explore in this chapter, this was the first realization of a full cellular automaton that uses a 2D grid by treating each cell as a finite state machine (FSM). An FSM is an entity that has a current state and changes state based on rules. As the name implies, a finite number of states and operations change the state of the cells, based on rules. Abstractly, FSMs are the building blocks of all modern deterministic computer programs.

The program we used to create our previous simple world, *SimWorld*, could be thought of as an FSM with a current state, representing which room the player's character is in. Certain commands, such as N or S, would change the state to be in another room, whereas other commands keep the state the same. You could define a table of states and transitions for *SimWorld* and implement it as an FSM. Von Neumann's cells had 29 possible states per cell, with hundreds of thousands of cells in total.

Although von Neumann's brilliant mind was pulled into other activities for the US government before he finished his cellular automata, his CA was popularized by John G. Kemeny in an article in *Scientific American* in 1955. Kemeny was also a legend in early computer science; he would go on to create the BASIC programming language, the first language for an entire generation of programmers (including me).

Von Neumann's goal wasn't to create cellular automata for any particular purpose other than to show it was possible to create artificial life – or at least one aspect of life that seems to be paramount: an entity that could reproduce itself, carrying information forward to future generations.

In the late 1960s, Edgar "Ted" Codd, while working for IBM, simplified von Neumann's states to eight instead of 29. This simplified the cellular automata by a lot. But it took Chris Langton, who also ended up at University of Michigan's cellular automata lab, to show how Codd's simplified model of cellular automata could be used to create a fully self-reproducing cellular automaton. Since each cell can have up to eight states, Langton designated these with a number (0–7), though today they are shown on computer screens with a color designating each of the states. His automata looked like a Q with small tail. The tail was a much smaller version of von Neumann's tape (without the original's requirement of 150,000 cells to work).

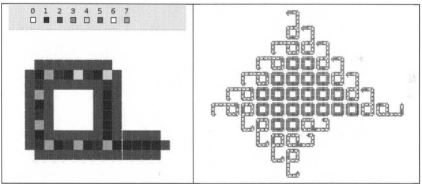

Figure 26: Langton's loop: a cellular automaton with 8 states and rules that allow it to reproduce itself.

The cells assembled into the Q with what look like outer layers and inner layers, not unlike layers you might see in a biological entity, as shown in the left side of Figure 26. In the inner layer was the information, or the DNA, that would be carried onto the next generation. By following a fixed set of rules, the tail extended with each step like an umbilical cord to a create new fetus, which detached, grew into a new version of the Q, albeit one with a little extra on the end of the tail, which served as a kind of umbilical cord to its parent Q.

Over time, this CA would produce a digital colony of organisms like that shown in the right side of Figure 26. The ones at the edge were vital and had the little Q tail, allowing them to reproduce, whereas the ones in the middle didn't have the tail or the extra information to pass on, so those were considered dead.

Langton's work was important not just because it was a simpler version of von Neumann's vision, but because it was a full realization of a self-reproducing automata in software, which had been von Neumann's goal all along. By the time Langton completed his loops in 1982, the idea of cellular automata had gained a life of its own.

The Best-Known CA: *The Game of Life*

The best-known CAs contain much simpler cells than von Neumann's 29 states or even Langton/Codd's eight states. In most CAs today, each cell has two simple states—dead or alive—which translates even better into binary values of zero or one and are well suited to be simulated via computers.

Perhaps the most of famous of these is *The Game of Life*, which was invented by Cambridge University's John Conway in 1970. Oddly, the idea of running this on a computer wasn't a part of the original formulation. Instead, being a mathematician at heart, he wondered whether some kind of simple cellular automata could be used not just to simulate artificial life but as a computing device. To illustrate his ideas, he created *The Game of Life* as a physical checkers-like board that lived in his lab on graph paper. He and his grad students updated the board manually, using a set of rules that he had devised, in which each cell (whose state was either alive or dead) changed its state based on the values of its neighbors. An example of patterns produced by *The Game of Life* is in Figure 27.

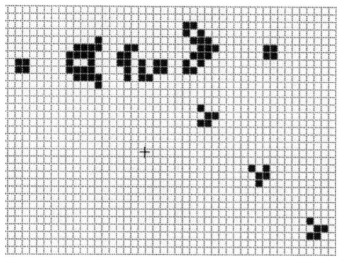

Figure 27: An example of some patterns in Conway's Game of Life *CA.*

The game begins a 2D grid that could extend indefinitely (in theory only; Conway, of course, had limited graph paper to work with). Whoever kicked off *The Game of Life* would give a particular initial configuration, which would mark some cells as alive (dark cells) while all the others were dead (light or white cells). The rules were fairly simple, if a little tedious to calculate, and like all cellular automata, were run in parallel for each step of the game.

You might say the game started at time $t=0$, and the cells recalculated themselves at $t=1$ based on the previous configuration. Each step in time (increase in t by 1) revealed a new configuration of the entire 2D grid. The rules (simplified from Conway's original) were:

- Any live cell with two or three live neighbors survives.
- Any dead cell with three live neighbors becomes a live cell.
- All other live cells die in the next generation. Similarly,

all other dead cells stay dead.

Different initial conditions produced interesting patterns on the game board. When run on a computer quickly, the steps looked like, well, microorganisms that were living and expanding and dying. These produced patterns are common enough that researchers have since classified them into basic categories: stable, oscillators, gliders, and chaotic.

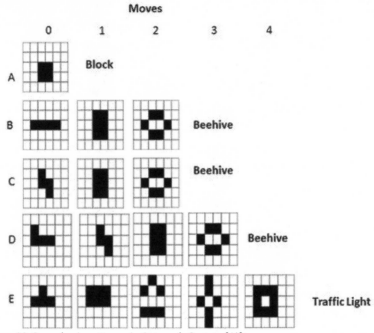

Figure 28: Some basic patterns in Conway's Game of Life.

The simple patterns are stable and oscillating. For example, in part A of Figure 28, you'll see that if you start with a block of four cells, that pattern doesn't change in subsequent steps. We say it is *stable*.

On the other hand, if you start with a simple pattern of four horizontal cells, as shown in part B of Figure 28 at time $t=0$, at the next step ($t=1$) you will see that following the rules results in a three-by-two vertical rectangle. If you keep going to $t=2$, you will see it become what's referred to as the stable beehive pattern. A that point, the beehive, like the block, is stable. If you go to $t=4$ or 5 or 101 or 10,001, it won't matter; the graphical pattern will stay the same.

One thing that makes the beehive interesting is that there are many ways to get to the beehive, but no matter how you got there, once you are there, you are stuck with that pattern (look at parts B, C, and D of Figure 28, all of which end up as a beehive).

There are also stable patterns in which cells that are alive will eventually die by following the rules. That means you end up with a board that has no cells lit up at all. For example, if you have two cells horizontally next to one another, following the rules will cause those cells to die, resulting in a stable blank board.

Perhaps more interesting are the oscillating patterns, in which two (or more) patterns repeat indefinitely. The Traffic Light pattern, shown in part E of Figure 28, is one such pattern. The rules flip the grid between the last two patterns. We say they oscillate indefinitely, like a traffic light.

It's important to realize that within complexity theory, an oscillating pattern is also stable in the sense that we get to a definite pattern that repeats. The shapes that repeat in an oscillator can get much more sophisticated as well, with names like blinkers and pulsars, each with a different period of repetition. Although these simple ones repeat with a period of two steps in time, it's possible to have patterns that repeat over a larger number of steps.

Perhaps even more interesting than the blinkers that oscillate are the gliders. The researchers at Cambridge called it that because if you follow the rules, the pattern, which is shown in Figure 29, reproduces itself over the course of a few moves, but it also erases its original cells. This means that over an interval of four time steps, the glider re-forms but glides to the right and below the cells of the original pattern. This means it has not just a period of repetition but also a displacement value. If you continue running steps in a board that has a glider pattern, the glider will eventually keep moving to the right (every four steps) and, at some point, disappear off the screen.

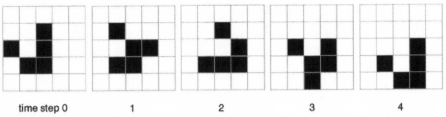

time step 0   1   2   3   4

Figure 29: An example of a glider pattern.

Although *The Game of Life* is a relatively simple program for today's computers (you can find it on websites and go through the game step by step or just let it animate and watch it), back then it was a new kind of software for computers.

CAs didn't become popular until Ed Fredkin, the brilliant but unorthodox director of the artificial intelligence lab at MIT in the 1970s who would later inspire Richard Feynman to propose the ideas behind quantum computers, took to them.

Fredkin, like von Neumann, believed that "the basis of life was clearly digital," [108] and he encouraged his group to dive into CAs, seeing parallels between DNA and the new little toy digital universes in cellular automata. During Fredkin's time, given that computers were slow and bulky, some members of the lab actually created a chip that was optimized for running CA quickly and could show the results on a monitor in the lab. This monitor left passersby mesmerized because it looked like there was some kind of organic life forms on the screen moving and evolving.

Wolfram and Elementary Cellular Automata

One of the biggest champions of using cellular automata to understand the world around us has been Stephen Wolfram, creator of the *Mathematica* software for symbolic mathematical computation. Wolfram, a brilliant if unorthodox physicist, attended Oxford at the age of 17 in the 1970s but found the lectures so boring that he left early to earn his PhD in theoretical physics from Caltech in 1980.

Wolfram became enamored with cellular automata when he saw that very simple rules could produce complexity that he wasn't expecting. He published a number of papers on the subject, becoming one of the pioneers in the field. In his own words:

> I took a sequence of simple programs and then systematically ran them to see how they behaved. And what I found—to my great surprise—was that despite the simplicity of their rules, the behavior of the programs was often far from simple. Indeed, even some of the very simplest programs that I looked at had behavior that was as complex as anything I had ever seen."[109]

From the point of view of cellular automata, Wolfram's studies focused on even simpler cellular automata than von Neumann's or Langton's loops, even simpler than Conway's *Game of Life*. These simpler automata allowed him to watch patterns develop graphically on the screen over time and led him to the conclusion that some processes were *computationally irreducible*:

> **Computationally irreducible:** Computations that cannot be sped up by means of any shortcut. [110]

In other words, computational irreducibility implied that the only way to figure out what would happen (in the future of a simulation) was to run through the computation and watch what happened. This is actually an important concept that we will rely on well beyond CA and may get into the purpose of running simulations in the first place.

Rather than using a two-dimensional grid like in *The Game of Life*, Wolfram focused on one-dimensional cellular automata. This would mean, basically, that the simple world of the CA consists of a single row of cells, and each of the cells in this row would have two states (not unlike the cells in *The Game of Life*, these cells could be on or off.) These were called 1D CA, or elementary cellular automata.

The rules in his 1D CA typically consisted of a cell deciding what to do in the next time step by looking at its two neighbors. This means that three cells would determine the fate of the current cell at each step: the cell to the left, cell in the middle, and the cell to the right. For example, if a rule said an alive cell with dead cells on either side of it would remain alive, you could express the rule in this way:

DEAD-ALIVE-DEAD → ALIVE

Therefore, if the left neighbor is dead, the current cell is alive and the right neighbor is dead; in the next step, the current cell should remain alive.

Unlike previous CA, which had a set of rules that was defined in a human language like English, Wolfram had the insight that you could define rules for elementary CAs by using a simple numbering scheme that relied on the binary representation of a number. You could then run different rule sets, using the same core algorithm, and draw conclusions about the outputs by specifying a different number (Rule #30 or Rule #90), for example.

It's possible that all of these types of elementary CA rules could be represented with a collection of eight bits. This is because there are only eight possible patterns of the three input cells needed for a single rule. Every single 1D automaton must then define what happens for each of these eight possible patterns; the only difference is whether to turn the cell on or off in the next step.

This means you can express the full rules for an elementary CA by using 8-bit numbers. The 8 bits are the values of the output cell for each of the eight possible patterns of input cells. Figure 30 shows a single rule set (a collection of eight individual patterns and an output rule for each one). Since 8 bits represent numbers between 0 and 255 in binary, this means there are 256 (or 2^8) rule sets, and you can refer to any particular rule set using the number represented by the binary of the output cells. In Figure 30, we see the binary value 00010110b, which represents rule set #22 (or simply rule #22 for short).

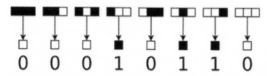

Figure 30: An example of a rule set consisting of 8 sub-rules.

One other aspect of 1D cellular automaton that makes it interesting is that now you can watch graphically what happens to the automaton over several time steps. This is done by still showing a 2D grid but having one row in the grid reflect one time step. The first row, row 0, would be the initial condition. The rule set is applied, and the second row has cells lighting up based on the applying the eight rules of the rule set on the previous row and so on.

Unlike *The Game of Life*, in which we were looking at one graphical pattern at one time step only (at a time), here you can visualize exactly what happens over time. We can see that graphical patterns emerge not just in space but in time. An example is shown in Figure 31: You can see the rule at the top and the pattern in the two-dimensional grid, starting with the top row at $t=0$ and each row below it representing one time step.

> **Time pattern** – a pattern which can be seen only by visualizing how a system behaves over multiple time steps. The pattern isn't visible if you just look at a system during one time step or two.

Now you see why running 1D (or elementary) cellular automata is interesting. Not only is it really convenient to show what happens graphically to the toy universe, you can see how it evolves over time in a single image, and patterns that only exist in time but laid out spatially (kind of like what physicists were trying to do with space-time diagrams).

In this rule (#90), a regularity emerges of triangles that you can see clearly. But you wouldn't be able to see these triangles if you looked at only one step in time or at a single row. To see the triangular pattern emerge, you have to look at it over both space and time. In real life or in a movie, for example, we usually only see one time step at a time, so this is a new way of thinking about things.

rule 90

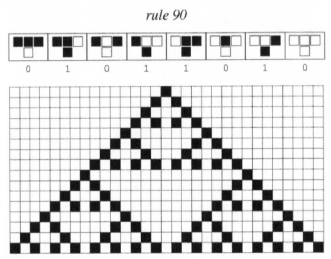

Figure 31: An example of running rule 90 in an elementary CA.

Reversible Cellular Automata

In *SimWorld* and other games, we talk often of loading a previous gamestate. In a cellular automaton (or in a graphical simulated world), you would have to store every single previous state to do this. For 1D CA, this is simple (depending on the number of cells), but a CA with more cells and that runs for a large number of steps (i.e., toward infinity), the information requirements grow toward infinity as well, which, as I've mentioned, is something that computer scientists hate. We immediately look for ways to optimize.

If it were possible simply to calculate the previous value of cells, we could go *back in time* in a CA by doing the same thing we did when we ran it forward. In other words, we would then have what's called a reversible CA, in which time can flow backward or forward with ease.

> **Reversibility:** In computer science, reversible computing is a model of computation in which the computational process is to some extent time-reversible.[111]

In a cellular automaton, this means that you could go backward in time and figure out the value of the previous row (or, more generally, the previous state) by using in some predefined rule set, just as you can move forward in time. We would be adding a form of deterministic time travel into CAs.

This was the question that confronted Tommaso Toffoli and Norman Margolus, whom Fredkin recruited to the MIT AI lab to work on CAs back in the 1980s. They wanted to know whether certain cellular automata were *reversible*.

It turned out that not all CA are automatically reversible. Although moving forward in an elementary CA is completely deterministic (i.e., there is only one possible value for the cell in the next step, based on the current status), this isn't necessarily true for all rule sets in reverse. In some rule sets, there are two possible previous values of a given cell (on and off), simply based upon the current values of the cell and its neighbor in the current row. These CA are said to be irreversible (at least without more information).

Toffoli and Margolus wrote a paper that showed that it was possible to construct rules in a cellular automaton that were reversible.[112] Although it showed that an arbitrary elementary CA was not reversible of its own accord, it was possible to determine whether a CA was reversible by using an algorithm, or a reversibility tester. *The Game of Life* wasn't reversible, for example.

Toffoli also found that it was possible to modify the rules of an existing CA to make it reversible. One way to achieve reversibility was to add rules that looked backward at the last few rows before deciding what value to put in the cell in the next time step. These are called second-order rules, because they rely on input not just from time t but also from time $t-1$ to decide the value at $t+1$. Toffoli also found that you could add a dimension—that is, if you made a 1D CA into a 2D CA with an extra bit for each cell, you could make it reversible.

Figure 32: Comparison of Rule 18 and Rule 18R, using second-order rules.

As an example, you can see rule #18 (for an elementary CA), which produces triangular patterns in time, modified to use a second-order rule, thereby becoming reversible (called Rule #18R). This also changes the time patterns that are being created, as shown in Figure 32.

For our purposes, let's remember why we are interested in reversibility. We are exploring the idea of a digital multiverse with multiple timelines. In a digital multiverse like this, time could be represented as a spatial dimension, where we can move forward or backwards in time. We could also, upon rewinding, change some variables before moving forward again, essentially creating new timelines.

Reversible cellular automata are interesting not just because they mean we don't have to store *every single gamestate at every moment in time*, which could conceivably turn into an infinite amount of information. They also give us the ability to rewind a graphical world, allowing it to go forward or backward in time just as easily.

As we scale up thinking about simulations from simple cases like *SimWorld* or elementary CAs, reversibility may be a key to building extremely complex simulations that rival the real world. You will see that I return to the concepts of reversibility and computational irreducibility as we start to look at more complex computations and simulations.

Simulating the Real World: Patterns and Fractals

Thus far, we have only looked at 1D and 2D CAs. What would a 3D CA look like? Well, it might look a bit like the natural world around us if the earth consisted only of flora and fauna. One of the reasons CAs and distributed rule sets of this kind (including fractals) are interesting is because they produce patterns not unlike those we see in nature.

One of the things that caused Wolfram to look further into CAs was that repeatedly applying rules created patterns that look like natural patterns. One of the elementary CA rules, rule #30, produced triangular patterns not unlike those you see on the surface of shells, as shown in Figure 33. However, while you can see the shell's pattern at a single point in time, the triangles in rule #30 can only be seen as time patterns by viewing multiple snapshots at the same time.

Figure 33: Rule 30 applied repeatedly produces triangular patterns that are similar to those found on certain shells. [113]

This brings us to the similarities between CAs and fractals, which are useful tools for understanding how the natural world works. Defined by Benoit Mandelbrot in 1975, a fractal is a mathematical and geometric concept whereby a shape has *self-similarity* at different scales.

Mandelbrot himself came up with the idea after trying to answer the question, "How long is the coast of Britain?" When he considered this question, he realized that the answer depends on what scale you are looking at. From a satellite image a coastline might look like a straight line that would give you one length, but if you actually walked every nook and cranny of that same coastline, you would wind up with a different length altogether. If you were an ant with a ruler, you might find another number, and so on.

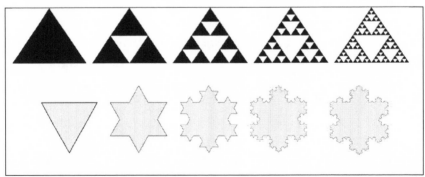

Figure 34: The Sierpinski Triangle and Koch Snowflake, examples of geometric fractals.[114]

Mandlebrot and others found that you could algorithmically start with simple shapes and repeatedly apply the same rule to them, resulting in self-similarity at almost infinite scales. Some common geometric fractals are shown in Figure 34; by taking the triangle on line 4, you can repeatedly apply the rule of "add an upside-down triangle of half length" in the middle of each dark triangle. If you repeat this over and over, you wind up with a shape called the Sierpinski gasket, which has some similarities to the triangles produced by rule #30 in the elementary CA shown earlier.

With this method, you can produce increasingly complex shapes that start to resemble nature. This is why fractal algorithms and chaos theory have proven valuable in the study of both periodic and nonperiodic phenomena in nature, ranging from error rates in electrical signals to the flow of turbulence (like that of smoke rising) to weather prediction. Within the video game world, fractal-type algorithms (i.e., algorithms with self-similarity) are also quite useful in generating lifelike images of flora and fauna in video games or computer simulations.

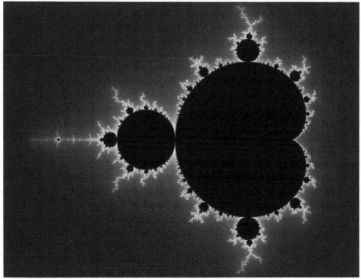

Figure 35: The Mandelbrot set.[115]

One of the best-known fractals is the Mandlebrot set, shown in Figure 35, which encapsulates self-similarity at many levels. If you look closely at the picture, you'll see that the same basic snowman-like shape is reproduced on each of the ancillary shapes. If you were to zoom in, you would find that each shape consists of ... more Mandlebrot shapes, ad infinitum, depending on how it was generated and how much resolution you could practically zoom into.

Like CAs, these geometric fractals are deterministic: you just follow the same rules again and again. In some ways, these fractals produce very stable outputs, at least in terms of what shows up at each scale. There are also random fractals, which rely on some level of probability embedded in a standard pattern. Some examples of problems in the real world that we can get insight into with random fractals include blood vessel simulation, polymer bonds, and even "ice floes drifting through the Bering Sea."[116] We'll talk a little bit about randomness within CAs shortly, but an example of the types of shapes you can form with random fractals include tree-like structures, formed from a random dendrite fractal pattern like that shown in Figure 36.

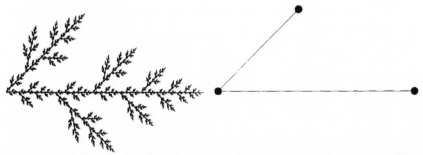

Figure 36: A random dendrite fractal pattern used to generate tree-like structures. [117]

CA, Fractals, Chaos, and Complexity

With CAs and with geometric fractals, we now have a simple, digital way to show how simple rules repeatedly run can create complex patterns in space (fractals) and complex patterns in time (CAs).

Going back to the results of a CA like *The Game of Life*, we found that some patterns were inherently stable, whereas some oscillated. How about the ones that didn't do either? Wolfram classified these different types of patterns into different categories—going from stable to complex to chaotic. Chaotic patterns were impossible to predict; the only way to figure out what's happening at step 1,942,000 would be to look at the step before it (step 1,941,999) and apply the rule set. This is why chaotic patterns are thought to be computationally irreducible. If you could figure it out without simulating that many steps because they settled into some predictable patterns, then the process is computationally reducible.

Chaos theory is all about studying what makes things irreducible; they are systems where small changes in initial conditions might have unpredictable results. These are called deterministic nonlinear systems more formally or, less formally, systems with *sensitivity* to initial conditions.

A commonly cited example is what's called the butterfly effect: when a butterfly flaps its wings in one part of this world (New York?) and it causes a storm in another part of the world (Hong Kong?), affecting the stock market in that part of the world. The butterfly effect is used because it is an eyebrow raiser, and it's impossible to predict all the small ripples that will be set off by an initial event without simulating the whole thing. Nevertheless, chaos theory, which is defined as a branch of complexity theory, assumes there are some rules or deterministic algorithms such that all these pieces are interrelated, like a very complex or chaotic cellular automaton.

One of the most famous chaotic problems is thought to be the three-body problem, first identified by Newton, named by French mathematicians in 1747,[118] and more recently made famous by

Chinese science fiction writer Cixin Liu's extremely popular book of the same name. If you have three heavenly bodies that are affecting each other completely deterministically (using Newton's laws of gravity), you can't say exactly what will happen after revolution 2 million without running through the calculations. This is usually expressed by giving the initial positions, mass, and velocity of the three bodies and asking whether they will settle into a stable orbit, or whether one or more of them will shoot off into some unpredictable behavior.

There are periodicities in nature, but there is also an element of chaos and complexity, which means that many deterministic methods will help us get there, but the only way to figure out what will actually happen is to simulate the natural world and watch. Moreover, we may need some element of randomness as well.

Multiple Timelines in CAs

If a CA is reversible, it means that the past is implicitly embedded within the present, because we can derive the past from the present. Like the equations of quantum mechanics, which seem to work whether we are moving forward or backward in time, a reversible CA works for delta $t=+1$ or delta $t=-1$.

Now, let's step back and see whether our dive into CA has added to our understanding of a simulated multiverse. The reason we would care about time travel and multiple timelines is so that we could, as Philip K. Dick would say, "alter the reality by changing some parameters," by going back in time and changing something.

What would two different timelines look like in a CA-type situation? Well, each timeline would be its own graphical pattern, so we couldn't really show two states at once without two different pictures side by side.

Since a timeline is more or less akin to a set of initial conditions and an update rule set, you could derive multiple timelines in an elementary CA by either changing initial conditions or changing the data of a row in the middle of a run (which means that the current row becomes the input of the new rule set).

You could also take the current row in the middle of a run of a CA, and from that time step forward, change update the rule to a different number.

Let's introduce the idea of multiple update rule CAs, so when the pattern is run, a cell can be arbitrarily changed, and the rule can continue to run, or we can select from one of several rules as a next step and we can see what the results might be. We can also rewind if we don't like the results, change the variables or the rules, and run it again.

If we could toggle between Rule 30 and Rule 99 at various points, who would do the toggling? It could be done by a random variable, but then we would no longer be in a deterministic set of rules, if it was clearly random. Of course, whether we can have a truly random rule using classical computing is an open question.

Stephen Wolfram discusses randomness in CAs as providing random input conditions. The randomness isn't integrated with the CA update rules; it is embedded in the run with the conditions in the first row. Different input conditions, even with the same update rule, might produce different patterns, though they generally tend to be very similar patterns. The randomness is put into the system by the person who presses Go or who provides the input data to the system. We now have a self-contained digital universe but we also have, just as we had with *SimWorld*, those watching the simulation and those affecting it.

Cellular Automata, Randomness, and Free Will

Some might object to my using CAs as an example of a framework on how to build simulations, particularly as an analog for being inside a simulation as complex as the world we live in, because a single update rule CA generally lacks both randomness and free will. Could we have a different type of randomness or even what we think of as free will inside cellular automata?

John Horgan examined this issue with respect to CAs in an essay in *Scientific American* after the death of Conway, the creator of *The Game of Life*.[119] He pointed out that Stephen Wolfram believed that CAs had free will, because the only way to find out what a CA will do, according to his definition of computational irreducibility is to watch it; it can't be predicted, which is completely compatible with the idea that the CAs have free will.

Conway himself argued that quantum processes have randomness, which implies that there is free will. Physicists, Conway argues, are free to measure or not measure the spin of particles, and there are dozens of ways to measure the spin, which determines, of course, what the results will be. This amounts to free will, argues Conway, on the part of the physicists as well as of the particle.

Is it possible to create a random CA? Wolfram and others have come up with a set of rules, based on values of different cells in the board at different points in the past, that creates a random sequence. This means that the probability that a cell will come up randomly with a value of 0 or 1 is about 50/50. This is a case when a deterministic rule set achieves randomness, at least as measured over time, but you could, of course, look at the formula and figure out what the value of a cell should be, so it's not completely random.[120] To get true randomness, though, we would probably need to rely on quantum techniques, and we'll revisit the idea of quantum computing and measurement in the next chapter.

Most scientists, certainly physicists, conflate the two ideas of randomness and free will. Horgan argues that free will, as we think of it, is about more: "They examine free will within the narrow, reductionistic framework of physics and mathematics, and they equate free will with randomness and unpredictability. My choices, at least important ones, are not random, and they are all too predictable, at least for those who know me."

In a sense, Horgan is arguing for an emergent definition of free will that is not reductionist, though coming at it from a more common-sense point of view. The higher-level reasons for doing something, argued Deutsch in *The Fabric of Reality*, are just as important as reductionistic rules and laws in determining what happens to the physical world. For example, he suggests, the laws of physics and chemistry might explain how bronze atoms stick together inside a statue, but they can't explain why the statue ends up in the middle of Trafalgar square in London. To truly understand that, you have to combine the low-level laws of chemistry and physics with higher-level, emergent properties—in this case, one of *purpose*. Whose purpose? That of the builders (the British parliament, for example, who might want to honor someone with a public statue).

What does this have to do with CAs and simulated realities? It actually brings us back the NPC versus RPG discussion: are there players who are having an impact on the game based on choices made by free will, or does the game just consist of the deterministic rules like in a CA?

These topics—randomness, free will, determinism and reversibility—are all related. We'll explore them through yet another lens in the next chapter, that of quantum computing, which brings these ideas together in a new kind of computer science.

Chapter 10

Quantum Computing and Quantum Parallelism

As it computes, the universe effortlessly spins out intricate and complex structures.

—Seth Lloyd, *Programming the Universe*[121]

Thus far in this part of the book, our discussion of how to build simulated universes, even toy ones, has relied on well-established concepts of classical computing. I am, of course, relying on a classical computer to write this book, which exists only as bits of information that my word processor renders on my screen to make it easier for me to edit. You are reading a product of those bits—either on a screen or on paper.

In this chapter, we'll look at how to compute a multiverse using the mechanism of the universe itself. To do so we will combine some of what we have learned about the quantum multiverse with what we have learned about bits and information on classical computers to talk about a new type of computing: quantum computing. This chapter serves only as a very high-level introduction to some ideas about quantum computing for the general reader, staying away the math, complex notations, and details of logic gates that a quantum programmer must master.[122]

My goal with this chapter is to explore how quantum

computing fits into our overall thesis about simulations and the multiverse. And it fits in quite well. In fact, it brings many of the concepts we've been exploring thus far—multiple timelines, parallel universes, and building tree-like structures that run forward and backward in time—together, like converging rivers of thought into a new, powerful intellectual stream that will carry us to parts IV and V of this book.

Modern Computers, Logic Gates, and Classical Computing

Before we move forward into the future of computing, let's go back a little bit. Whereas modern computing is based on bits, the logic followed by those bits pre-dates modern computers by many decades. In fact, the logic that underlies many of our ideas of computing was developed as a mathematical formulation by George Bool in the 1800s, well before electrical circuits, let alone digital computing.

Boolean logic (more formally called Boolean algebra) is a branch of algebra that defines a limited set of special operators that act on variables. Unlike regular algebra, where variables can take on any value, Boolean variables can only have one of two possible values: TRUE or FALSE. Boolean algebra has operators just like regular algebra (whose operators include add, subtract, multiply, etc.). The most common Boolean operators include AND, OR, and NOT, as shown in Figure 37.

As an example, the expression (x AND y) is one way to represent an operation (AND) on two input variables (x and y); the output is TRUE if both variables are TRUE, but FALSE otherwise. The NOT operation changes the value of the input (from TRUE to FALSE or vice versa); the OR operation returns TRUE if either of its input variables is true and FALSE if neither is true.

For forever linking Boolean logic to electronic circuits (and later to information theory), we can lay some of the blame on Claude Shannon, whom many consider the father of information theory. While Shannon was getting his master's degree at MIT in 1937 (more than half a century after Bool created Boolean algebra), he was asked to find a way to simplify telephone switches, which were built with lots of wires in messy arrangements. Telephones had grown in popularity, becoming the hot new technology of the early twentieth century (much like smartphones a century later), resulting in horribly complicated networks and switches, which were expensive to debug and maintain.

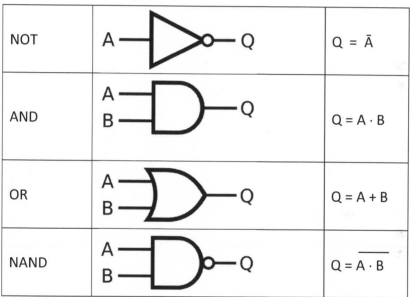

Figure 37: Boolean logic gates usually take one or two inputs in binary and output one value.

Shannon realized that physical switches and relays, which were used to connect telephone circuits, were acting in a way analogous to bits of information. A value of zero (or FALSE) meant that no signal was going through the switch, while a one (or TRUE) value meant that a signal was present.

Shannon's insight was that the properties of Boolean logic

(using operators like AND, OR, NOT) could be applied to the electrical circuits, greatly simplifying how switches and circuits were put together. These became known as logic gates, because they were electrical components that let through signals based on Boolean logic. You could use logic gates to design circuits that would be much easier to build upon, debug, and modify.[123]

Logic gates are still the building blocks for most of our computers today; in the 1960s it was realized that silicon-based transistors could implement these gates more efficiently and compactly, resulting in microprocessors, consisting of large numbers of transistors. These logic-based microprocessors not only powered the first personal computers like the Apple II and Commodore 64 that I used as a kid, but they continue to power our smartphones today.

Why is this discussion relevant to quantum computing? Because before we had high-level programming languages like BASIC and FORTRAN (leading to popular languages like JavaScript and Python today), programmers had to think about logic gates and how they would be combined in operations to write low-level *programs* that could best be thought of as logic circuits.

Today's quantum computers represent perhaps the first truly new type of computing since the 1950s. Although still based on bits and logic gates, we need to learn about *quantum bits* and *quantum logic gates* to understand today's quantum computers. And, in the absence of high-level programming languages, we have to think in terms of combining quantum logic gates to build quantum circuits in the same way that Shannon and other pioneers of modern digital computing had to create circuits that combined logic gates. In 50 or 100 years, when quantum computers are more widespread, they might look back at the 2020s (when this book was written) as the dark ages of quantum programming when we still needed to learn about quantum gates and circuits!

The Origin of Quantum Computing

The history of quantum computers is an interesting bridge between the history of computation and quantum physics. In fact, many of the early players in quantum computing were physicists who decided to focus on computer science (or vice versa).

The man most credited with the idea of quantum computers was Nobel prize-winning physicist Richard Feynman, who wrote a now-famous paper in 1981, called "Simulating Physics with a Computer."[124] In it, Feynman credited his interest in simulating quantum mechanical processes on computers to Ed Fredkin (the same one who ran the cellular automata group at MIT that we discussed in *Chapter 9: Simulation, Automata, and Chaos.*

Feynman supposed that existing classical computers could approximate physical phenomena by using differential equations. But to truly simulate what a set of quantum particles would do, we would need to simulate all of their possible values and interactions. To do this, you would need to keep track of what particles were doing in different parallel universes and how they were all interfering with one another (which is, according to the MWI, what Schrödinger's wave equation is all about).

By simulating quantum processes in this way, Feynman reasoned that we could solve problems that were extremely difficult for traditional computers. It turns out that approximating physical phenomena like molecules is one of a class of many problems that grow exponentially and thus are referred to as *intractable* using classical algorithms.

A famous example of a problem that grows exponentially is given in the old Indian story of a king who liked to play chess, and a sage who tricked him. The sage's reward if he beat the king was to be one grain of rice on the first chess square, two on the second, doubling the grains of rice until the end of the chessboard (64 squares total). Unbeknownst to the king when he agreed to play, it turned out the king would need to give him 2^{64}, or 18,000,000,000,000,000,000,000 grains of rice (roughly equal to 210 billion tons of rice), which would take more space than available in all of India![125]

Other problems that grow exponentially like this include breaking modern encryption and cryptography, drug discovery (by simulating different molecules and their reactions), optimizing logistics and operations, and simulating complex financial markets, all of which quantum computers are expected to be able to help with. These types of intractable problems that grow exponentially would take a standard classical computer, like the one I'm using now to write this book, *thousands of years* or more to crack, depending on the number of bits in the encryption key. Modern cryptography is built not on the idea that codes can't be broken, but that it would take a classical computer way too long to crack a single code, rendering it impractical.

Simulating aspects of the real world that are intractable is one of the big promises of quantum computing and what Feynman was considering when he wrote about the concept of a quantum computing simulator, which is the predecessor to what we call quantum computers today.

Some Characteristics of Quantum Computers

Although I had heard of Feynman's lectures and paper long ago (it's referred to in almost every article or book about quantum computing), it wasn't until I read his papers again while writing this book that I noticed both the connections to cellular automata and time, as well as simulation theory in general. Though Feynman's theoretical work pre-dates any actual quantum computer, he brought up several important concepts that relate to the idea of quantum computing, and these will tie into our overall thesis of the multiverse as a computational structure:

Discrete time. One issue that Feynman brought up in his original paper was how to simulate time. He suggested that we assume time is discrete, particularly since we cannot measure it practically at less than some threshold value (he suggested it was 10^{-21} but even if it was smaller, say 10^{-41}, that wouldn't matter, as long as it was discrete). He then referenced how cellular automata represent time by going from step to step (which we saw in Chapter 9) or by space-time diagrams where the time axis has discrete markings on it, and how this was all very similar to how computers work.

Logic gates. In his follow-on paper from 1985, "Quantum Mechanical Computers," Feynman spent quite a bit of time discussing both classic logic gates and how they might work in a proposed quantum computer. In fact, most programming in quantum computers today involves understanding quantum logic gates to some level.[126]

Reversibility. He references one more aspect of quantum computing that is important both to the idea of quantum computers and to our topic in this book: reversibility. A classical computing operation isn't necessarily reversible. if you have the result of an AND gate as 0, you don't know whether the inputs were (0,0), (1,0), or (0,1). But quantum mechanical phenomena are thought to be reversible; then the quantum computer would also need to be reversible. He specifically referenced the reversible cellular automata of Fredkin and Toffoli at MIT in his paper.

Probability and interference. David Deutsch, whose work we mentioned in Chapter 7, was an Oxford physicist who became one of the pioneers of quantum computing. Deutsch explains in *The Fabric of Reality* that Feynman saw that by using particles, like photons, for example, as the basis for storing quantum information, you would have to deal with the probabilistic nature of quantum phenomena. This means that you'd have to compute every single probability to figure out what is going on. This would take a traditional computer a very long time. However, the beauty of the quantum multiverse is that a single particle already is computing all of these possibilities—in every single parallel universe! So Feynman realized that if you could leverage the phenomenon of interference, you could sum up across all the versions of the particle in the multiverse (or, using the Copenhagen interpretation, sum up across all the probable versions of the particle) and come up with an answer to your problem, which is actually how quantum computers work.

Qubits, Parallel Universes, and Computation

The idea of quantum computing started to take off in the 1990s as computer scientists and mathematicians began to study how you could solve problems by using this revolutionary approach to computing. Much of this research was theoretical in what was considered an obscure corner at the intersection of computer science, mathematics, and quantum physics. Much like George Bool's Boolean algebra, there were no computers that could implement quantum algorithms, so it was all theoretical.

Eyebrows were raised in 1994 when Peter Shor, working at Bell Labs, came up with a quantum algorithm that could break most modern encryption by using quantum computing algorithms. Today's encryption is based on the difficulty of factoring large numbers. Even today, although there are no quantum computers that can implement Shor's algorithm in full yet, there is worry that most of our encryption will be broken in a few years as more capable quantum computers come along. When this happens, there will be a rush to quantum-safe encryption algorithms (which cannot be broken quickly by either classic or quantum computers).

Just as classical computers operate on bits (which can only have a value of 0 or 1), the fundamental unit of quantum computing is the *qubit*. A qubit is a classical bit that has, for lack of a better analogy, gotten drunk and can't decide whether its value should be 0 or 1. This is done by putting the qubit into a state of *superposition*.

Unlike a classical bit, a qubit doesn't have to be in a definite state; rather, it can exist in a superset of the possible states. You'll recall we discussed superposition of particles in Chapter 5. Since a classical bit has only two possible values (0 or 1), a superposition is the superset of all of those possible values, and represented like this: {0,1}.

A qubit could have *either* or *both* of those values until it's measured, just like an electron or photon can be in a superposition until it's measured. This is similar to the absurdity of Schrödinger's cat being both alive and dead until someone opens the box and takes a look to observe the state of the poor cat, or the idea that a photon goes through both slits until it is measured to go through just one.

So how can a qubit be used in any kind of computation if it's not in any definite state? The idea is that by allowing it to be in superposition, you can compute all the possible values of any computation involving that qubit, each of which *theoretically occurs in a parallel universe*. Although you don't need parallel universes strictly to talk about quantum states, it is a very useful way to visualize how a quantum computer works, which can seem like magic otherwise.

This idea that a quantum computer can speed up exponential problems by computing every possible value in a parallel world is referred to as *quantum parallelism*.

Julian Brown, in *The Quest for the Quantum Computer*, describes a conversation with Deutsch, who is not only a big proponent of the multiverse interpretation of quantum mechanics, but is not shy about applying it to quantum computing: "If an electron can explore many different routes simultaneously, then a computer should be able to calculate along many different pathways simultaneously too."[127]

Deutsch points out that that to factorize 250-digit numbers means that a computer would have to search 10^{500} possibilities, which would be intractable for normal computers. Shor's algorithm can do it much more quickly. Deutsch makes the argument that since there are only 10^{80} atoms in the entire visible universe, there is no way we would be able to compute 10^{500} possibilities in any reasonable amount of time with a classical computer.

So, how is it possible that a quantum computer can do this? Deutsch says that the only way such a calculation is possible is by having 10^{500} universes calculate simultaneously, each calculating one set of values, and then retrieving the answer back to our universe.

If we think of calculating using a single bit (or rather, a single qubit), since there are only two possibilities, we would need to only compute possible values. You don't really need parallel universes to compute the two possibilities to figure out the best answer. But as we scale the number of qubits, say to a *qubyte* (8 qubits), then there are 2^8 (or 256) possible values of those bits. If we wanted to explore all of these 256 possible values (ranging from 00000000b to 11111111b), we would theoretically have to compute based on every single possible value of the bits, one by one. Even 256 is not a very big number, and you could be deceived, like the king with the chessboard was. As you scale the number of bits to 64, like the king who owed the sage 2^{64} grains of rice, the requirements rise exponentially; you would need to run each of those values through your computation and get a result in serial, a process that would take many years.

On the other hand, if we were to use a quantum computer, we could compute each of these in a separate universe, and sum up the results so we could observe the most ideal answer, using interference and quantum measurement, in a very short period of time.

Creating Qubits and Quantum Computers Today

The main challenge in creating quantum computers was to use physical objects (atoms, molecules, photons) to represent qubits in a reliable way. This was, of course, a challenge for storing regular bits in classical computers too, back in the day. It was solved by using certain voltages to represent a 1 and certain other voltages to represent a 0. Later, this was optimized to use light in fiber optics to represent 0s and 1s. These techniques don't really work for quantum computers, which is why it's been difficult to create large-scale quantum computers.

We've known for a long time that every physical particle contains information (such as its spin); the key was to find a way to represent this in a physical computer and keep track of these values. The main problem today is that qubits are unreliable: *any interaction* with another particle in the physical universe could change the value of a qubit, which would throw the whole calculation off. It's difficult to isolate the particle that represents the qubit so they don't randomly produce errors.

In quantum computing systems, to make sure the values are reliably stored, *entanglement* is often used across multiple particles. This means that all the particles have a value that is related to the other particles, so we can reliably figure out what the "correct" value of the qubit is. If a particle *decohered* by taking on an erroneous value, or been measured to take on a specific value prematurely then the system can figure out that something erroneous happened. This can be done using standard error detection techniques which are used today in transmitting bits over the internet or over wireless transmission, to ensure that the bits at the destination are the ones that were sent from the source.

This means that most quantum computers require many particles to represent one qubit. Moreover, when you run a quantum algorithm, you have to run it many times to be sure the value is the correct one and no errors were introduced into the qubits during the run; running the same program 100 times to ensure the right answer isn't unusual. Still, this is much faster than reverting to classical computers to solve intractable problems.

The difficulty of maintaining qubits (which often require superconductors to cool the particles) is why we only have primitive, room-sized quantum computers today that are run from the cloud. These are created mostly by giants like IBM, Google, Microsoft, and Amazon, as well as from startups like D-Wave, which have a very limited number of qubits. To solve the encryption problem, we would need reliable 256-qubit quantum computers, which haven't appeared as yet, though every year we are getting closer.

Figure 38: The IBM quantum composer user interface

An example of IBM's cloud interface for programming 5 qubits is shown in Figure 38. It is only a matter of time before these companies figure out how to keep track of qubits at room temperature, and we might see an explosion of quantum computing power at that time, perhaps even personal quantum computers.

Algorithms Using Quantum Computing

Now, stepping away from the hardware, how do we compute using quantum computers? This is where quantum parallelism, as Deutsch calls it, comes in and we start to be concerned with not only what is happening in our universe but in all possible parallel worlds. When we have a problem to solve, such as looking for the factors of a number, many possible values must be tested. It simply takes too long to test every single number—at least in one single universe.

> **Quantum parallelism:** The ability of a quantum algorithm to run a function for all possible values of input variables at the same time by putting those variables into superposition. This is the primary advantage of quantum computers/algorithms over classic computers/algorithms, which can only run a function once at a time for each possible input value.[128]

You'll notice that in the formal definition I didn't specifically say parallel universes, though the implication of a value being in superposition is there—that it has all possible values. The idea is that one or more of the different values of the qubits (in one or more of the parallel universes, if you subscribe to the many worlds interpretation of quantum mechanics) will give us the right answer.

But how do we get an answer from a quantum computer? In the same way that a particle needs to be measured to get to a single possibility, the same happens with a qubit. Once it is measured, that bit is no longer in superposition and takes on a definite value of 0 or 1.

The simplest way to implement a quantum algorithm would be to (expressed in a classic computing sense):

1. Arrange a set of inputs (qubits) that represent one set of input values that need to be computed.

2. Implement the computation, using (quantum) logic gates in a circuit.

3. Make sure the circuit puts the qubits into superposition that we need to run in parallel.

4. Arrange an output qubit(s) that tells us the answer we are searching for when we are ready to measure.

Actual quantum computing is more complex, as you might imagine. Designing an algorithm using quantum computing requires thinking not just about the circuit to do the computation, but also how best to measure the output. Figure 39 shows an example of a logic circuit for Shor's algorithm, with the blocks representing either individual gates or collections of gates.

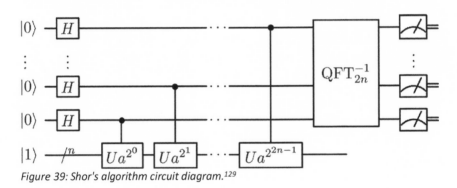

Figure 39: Shor's algorithm circuit diagram.[129]

Logic Gates in Quantum Computing

In the same way that AND, NOT, and OR gates were used to create the infrastructure for classic computing, quantum gates have become the way to think about programming quantum computers.

The idea of these gates was well specified even before we were able to build quantum computers. Although higher-level languages are now emerging for programming quantum computers, building such programs still requires an understanding of quantum logic gates and circuits.

In Figure 38 (the IBM cloud interface for programming its quantum computer), you can see the lines that represent each of the qubits and the gates represented as boxes. A programming language in the bottom right looks like a normal programming language, but for the computer to execute the program, it has to be translated into a quantum circuit with qubits.

Operator	Gate(s)	Matrix
Pauli-X (X)	—X— —⊕—	$\begin{bmatrix} 0 & 1 \\ 1 & 0 \end{bmatrix}$
Pauli-Y (Y)	—Y—	$\begin{bmatrix} 0 & -i \\ i & 0 \end{bmatrix}$
Pauli-Z (Z)	—Z—	$\begin{bmatrix} 1 & 0 \\ 0 & -1 \end{bmatrix}$
Hadamard (H)	—H—	$\frac{1}{\sqrt{2}} \begin{bmatrix} 1 & 1 \\ 1 & -1 \end{bmatrix}$
Phase (S, P)	—S—	$\begin{bmatrix} 1 & 0 \\ 0 & i \end{bmatrix}$
$\pi/8$ (T)	—T—	$\begin{bmatrix} 1 & 0 \\ 0 & e^{i\pi/4} \end{bmatrix}$
Controlled Not (CNOT, CX)		$\begin{bmatrix} 1 & 0 & 0 & 0 \\ 0 & 1 & 0 & 0 \\ 0 & 0 & 0 & 1 \\ 0 & 0 & 1 & 0 \end{bmatrix}$
Controlled Z (CZ)		$\begin{bmatrix} 1 & 0 & 0 & 0 \\ 0 & 1 & 0 & 0 \\ 0 & 0 & 1 & 0 \\ 0 & 0 & 0 & -1 \end{bmatrix}$
SWAP		$\begin{bmatrix} 1 & 0 & 0 & 0 \\ 0 & 0 & 1 & 0 \\ 0 & 1 & 0 & 0 \\ 0 & 0 & 0 & 1 \end{bmatrix}$
Toffoli (CCNOT, CCX, TOFF)		$\begin{bmatrix} 1 & 0 & 0 & 0 & 0 & 0 & 0 & 0 \\ 0 & 1 & 0 & 0 & 0 & 0 & 0 & 0 \\ 0 & 0 & 1 & 0 & 0 & 0 & 0 & 0 \\ 0 & 0 & 0 & 1 & 0 & 0 & 0 & 0 \\ 0 & 0 & 0 & 0 & 1 & 0 & 0 & 0 \\ 0 & 0 & 0 & 0 & 0 & 1 & 0 & 0 \\ 0 & 0 & 0 & 0 & 0 & 0 & 0 & 1 \\ 0 & 0 & 0 & 0 & 0 & 0 & 1 & 0 \end{bmatrix}$

Figure 40: Popular quantum computing logic gates.[130]

Gates in quantum computing can be thought of as analogous to classic gates but with some important differences. Rather than being named simply after Boolean logic, these gates are often named after a person, usually the one who described how the gate could work on a physical level, and are usually represented by a letter. Some of the common gates are shown in Figure 40.

The intricacies of these gates are beyond the scope of this book, since we are interested in the relationship between quantum computing and the simulated multiverse. Suffice it to say that like classic logic gates, a quantum gate gets one or more qubits as input and operates based on particular logic, outputting certain values. The input and output values can be represented in a table of inputs or outputs, like with classical gates or, for those more mathematical, in a matrix that is applied to the input values.

As an example, the Pauli X gate (represented with an X in a box) reverses the value of a bit from 0 to 1, so it acts like a classical NOT gate with one input and one output. The X gate and several other quantum gates (Y and Z) are named after the German physicist, Wolfgang Pauli, who discovered the Pauli exclusion principle.

For a gate that takes more than one input value, such as the Controlled Z or SWAP or CNOT gate (called multibit gates), the qubits that are sent in are linked (using quantum entanglement), which is represented by a vertical line connecting the two qubits in a circuit diagram.

Perhaps one of the most interesting (and important) quantum gates is the Hadamard gate, or *H-gate*.[131] The H-gate transforms a qubit of any value (remember that a qubit can have a value of 0 or 1 or an indeterminate value of both 0 and 1) into superposition. At that point, theoretically at least, we are back in the problem of quantum indeterminacy and Schrödinger's cat: the cat is both alive and dead until a measurement is made. In the MWI, this means there are multiple parallel universes with each value. If we have multiple qubits in superposition, say 8 bits, we have 2^8 (or 256) multiverses, and so on.

Another interesting gate is the CCNOT gate. If you look at the table of inputs and outputs for this gate, shown in Figure 41 you'll see that it operates just like an AND gate for the first two bits if the third input bit is 0; otherwise, it acts like a NAND gate (a combination of a NOT and an AND gate applied in sequence) if the third bit is set. Unlike a classic gate, which takes two values and gives one output bit, this gate takes three inputs and has three outputs, changing the values based on the logic of the gate.

The interesting thing about the NAND gate is that it is universal for classic computing operations. You could implement any computer by using just this one gate (though you would need lots and lots of NAND gates, and it would be very slow). Similarly, the CCNOT gate is universal, too; it could be used to implement any classic computation on a quantum computer (though it would be quite inefficient and require a lot of gates).

Inputs a b c	Outputs a′ b′ c′	Circuit Diagram
0 0 0	0 0 0	
0 0 1	0 0 1	
0 1 0	0 1 0	
0 1 1	0 1 1	
1 0 0	1 0 0	
1 0 1	1 0 1	
1 1 0	1 1 1	
1 1 1	1 1 0	

Figure 41: Input-output table and circuit diagram for Reversible Toffoli Gate (CCNOT).[132]

More important, this gate is called the reversible Toffoli gate, for reasons we will get into, for a good reason. Tommaso Toffoli, whom we mentioned in Chapter 9, was best known for his work on reversible cellular automata and reversible logic gates.

Time Travel and Quantum Reversibility

In 2019, a curious article came out in the *New York Times*; its headline said, "For a split second, a quantum computer made history go backwards."[133] It was an overview of a paper by a group of physicists in Moscow, Europe, and the United States, who had succeeded in getting a qubit to go back to its previous state. It wasn't really time travel per se—or was it? What we have been exploring in the idea of a digital or simulated universe is that time travel is the equivalent of bits of information being sent back to the state they were in at some previous time.

They succeeded in getting several entangled qubits, through electrical excitement, to go back to their previous state just a fraction of a second earlier. This only worked 85% of the time with two qubits and only 50% of the time with three qubits. The degradation of performance was blamed on decoherence of the particles because of errors and interaction with the environment.

Of course, for it really to be time travel, you would have to make *all of the bits of information* that are being rendered go back to the way they were at some previous time. If this were done for a simulation like ours, to go back to, say, 1961, that looks like a herculean task compared to what this team accomplished.

Still, could quantum computers provide for a way to do this? All quantum gates are thought to be unitary and reversible. By unitary, this means that all the probabilities must add up to 1. In a multiverse model, it means that the probabilities of each of the different states of the particle, when interfering with each other, must add up to 100%. Quantum gates are thought to be reversible because in theory the equations of quantum mechanics work forward and backward.

Let's revisit the idea of reversible gates for digital computers. A simple AND gate is not considered reversible, because we don't know from the value at the output what the inputs were. Controlled NOT gates are more easily reversible, and the CCNOT gate, or Toffoli gate, is known for being reversible.

Based on the three outputs, you can figure out exactly what the input values were. This means that not only can the Toffoli gate be used for universal computing, but it raises the specter of quantum computers not only simulating the universe forward but also of simulating the universe backward, thus fulfilling Feynman's vision of a quantum computing simulator.

However, there is a catch, and that is quantum measurement. Quantum measurement by itself is not thought to be reversible but is one of the few things in physics that is truly random. Once you perform a measurement of a qubit and it takes on a definite value of 0 or 1, you cannot go back to the exact state it was in before it turned to superposition.[134] This means that quantum measurement is a truly random process: you cannot predict which value will come up with any certainty, and you cannot predict what the value was without some additional information.

Therefore, all quantum logic gates are reversible, but quantum measurement is not. This seems like a strange distinction but one that means we can reverse a quantum circuit at any point except at the end, which is when measurement is done. Still, the reversibility of quantum logic gates gives us the sense that there may be ways to move back in time without having to load previous gamestates in a digital quantum multiverse. Of course, our classical Core Loop algorithm doesn't rely on reversibility per se; it relies on saving and reloading gamestates, but this gives us a much more efficient way (in terms of storage, though not necessarily performance) to achieve the same effect.

The Core Loop and Quantum Computing

Before we tie the world of quantum computing to our Core Loop, let's recap what we have learned about quantum computing. The basic premise of quantum computing is that the universe already computes; its fundamental building blocks, whether atoms or electrons or photons, already hold information, and this means the universe can be used to compute.

David Deutsch goes much further and ascribes the power of quantum computing to its ability to operate across all of the possibilities simultaneously. A quantum computer is, in essence, a computer with feet across the multiverse, and it can draw on this computational power to complete difficult (more specifically, intractable) problems in short order. This is done by putting a number of qubits in superposition so that each one has both a 0 and a 1, which automatically creates 2^n possibilities (where n is the number of qubits).

Rather than having to process what would be an exponential number of possibilities in serial (a process that could take literally thousands of years for intractable problems), a quantum computer harnesses the fact that each bit relies on a single quantum element, which has many shadow elements across the multiverse already, operating in parallel. Deutsch argues that quantum parallelism and is not just an aspect of quantum computers; it is an aspect of the universe that we live in (or rather, more accurately, the multiverse that we live in).

We then use the fact that these particular parallel universes spawned out of this one—the one asking the problems—and then we use interference across these universes as a way for them to interact with one another. Through quantum measurement, we can find the answer to the problem or, rather, find a single answer to the problem out of the many answers that might exist in all possible universes.

If we were to diagram this out, we find that what is happening when we give a quantum computer a task is very similar to our Core Loop, except that we can now rely on parallel processing rather than on a single computer in a classic sense. Figure 42 shows this process of branching out to many possibilities, computing an answer in each one, and then getting the best answer. This figure uses letters to represent the state of each specific step in the computation. Each step is like the gamestates we explored in *Chapter 8: Multiple Timelines in SimWorld.*

In Figure 42, the philosophical implications of what might be happening here aren't quite settled. Are we somehow jumping to the parallel universe that has the right answer? Or are we simply navigating our consciousness to that parallel universe? Or are we merging it with the one that we are in now? Or is something else entirely going on?

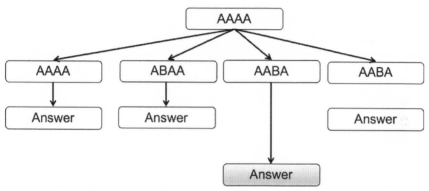

Figure 42: A diagram of quantum parallelism at work.

Another big question is what happens to the other universes, the ones that don't have the right answer for our computation? I would argue that from the perspective of the quantum computer, those are no longer needed for the computation and are essentially discarded. Or sitting in memory ready for garbage collection in the world of multiverse computers. Others might argue that those universes continue to exist, and we get an almost infinite number of physical universes. We see science once again reverting to the magic of infinity, which basically allows for anything and everything.

So why did I spend all this time going into the details of quantum computing? It was to give you a flavor of qubits and a quantum computation, but really it was to show how in one way of looking at it, quantum computing seems to be following our Core Loop. It seems quantum computing is just like the quantum multiverse branching off processes in space and time and then allowing us to optimize some computing to get some outcome and then merging to discard the unneeded computational worlds.

In a sense, a quantum computation algorithm is all about searching different possibilities, each of which can be thought of as a separate gamestate of a simulated universe and finding the one that has the best answer.

The basic argument that I'm exploring in this book is that perhaps something like this is already going on all around us, and that it is a natural process: we live in a computational multiverse. The qubits are all of the elementary particles of our universe, arranged together in some kind of program to compute *something. This something that we call our universe is some kind of computation, and the ability for particles to hold information is key to this computation.*

Just as John Wheeler said *"it from bit,"* Deutsch has updated this, using the idea of quantum computers: *"it from qubit,"* realizing that the universe consists not just of bits but of quantum bits that can represent all possibilities as part of an ongoing computation.

With this kind of computation, we now have a way to explain a mechanism for ideas that sounded just like science fiction referring back to Philip K. Dick and the Mandela Effect.

In an attempt to formalize this a little more, we will explore two core concepts in this book, the Multiverse Graph and the Core Loop, in the next part of this book, before finally stepping back in the last part to ask big questions about what it all means.

DEVS AND RE-CREATING THE PAST

In the Hulu series *Devs*, released in 2020, we follow Lily as she tries to unravel the mysterious death of her boyfriend. The boyfriend, who, along with Lily, also worked for the mysterious San Francisco Bay area tech giant Amaya, had recently joined the ultra-secret Devs group. Only a few people know what is going in Devs, one of them being Forest, the CEO of the company, who has been emotionally disturbed since the death of his wife and his daughter Amaya, after whom the company is named (not to mention the creepy giant adolescent statue of Amaya that serves as the company's logo).

It turns out that Devs is basically a quantum computer that is isolated from the outside the world. This quantum computer combines the idea of classical determinism with quantum algorithms. Throughout the first season, we learn that the immediate task is to try to re-create what happened at any time in the past. Theoretically, they could go back to the time of Jesus and watch what happened. Underlying this is the idea that we are all on tramlines that extend to the past and the future. The program isn't quite working until one of the developers introduces Everett's many-worlds ideas, and then voila! They are able not just to hear but accurately to see what happened with Jesus dies on the cross. They can also predict exactly what will happen in the future by moving this simulation that they have created on a quantum computer (QC) backward and forward. I won't reveal exactly what happened at the end of season 1, but it involves going back in time in the simulation to a previous point and then moving forward from there. Sort of.

The show brings up not only quantum computers and the multiverse interpretation but also the issue of free will and whether the future is completely determined by the past. Since QCs should be reversible, you can travel to any point in the past and restart the program there.

Part IV

Algorithms for the Multiverse

"Have you also learned that secret from the river; that there is no such thing as time? That the river is everywhere at the same time, at the source and at the mouth, at the waterfall, at the ferry, at the current, in the ocean and in the mountains, everywhere and that the present only exists for it, not the shadow of the past nor the shadow of the future."

— **Hermann Hesse,** Siddhartha

Chapter 11

Digital Timelines and Multiverse Graphs

In this chapter and the next, we will get into more details about two items I have been mentioning throughout the book, and which hat will help us visualize how a simulated multiverse might work: the Multiverse Graph and the Core Loop. Although multiple timelines have become common in science fiction, and physicists have made talking about a multiverse of parallel universes acceptable, there really hasn't been a general agreement on how to represent multiple universes and timelines in a graphical way so that we can intuitively grasp what's going on with a glance. This includes not just how to represent the multiverse but how to traverse this graph as time marches forward.

Casual readers who are not interested in the details can feel free to skim these chapters and skip to the last part of the book, where we'll transition to take a broader view of the nature and purpose of the computation of our simulated multiverse, and finally, I'll end the book with a chapter on what it might mean for us, the inhabitants of a complex multiverse, from a spiritual perspective.

Representing Timelines

Thus far in this book, we've included a number of visual representations of timelines, ranging from standard Minkowski space-time diagrams to cellular automata showing a graphical world evolving on a single timeline, to diagrams with tree-like structures for showing multiple timelines. We have also looked at digital ways of representing a gamestate, or state of the world, which encapsulates some amount of information into bits (either regular bits or qubits), which might represent a point in the tree.

When discussing multiple timelines and a multiverse, it's only natural to think about tree-like structures, since worlds are branching off from what looks to us like a main timeline. Realistically, though, if we live in a multiverse, there is no main timeline, though to us it might look as if we were on (using the *Arrowverse* terminology) Earth 1.

In addition to thinking of timelines as multiple possible futures, with the introduction of quantum phenomena like the delayed-choice experiment and the weirdness of the Mandela Effect, we explored the idea that there may be multiple pasts or, as Schrödinger first referred to them, multiple simultaneous histories. This is typically represented in an inverted, tree-like structure, more like rivers merging than like a tree branching out.

Thus, any graphical representation of this process needs not only to account for branching but also for merging of timelines. Simple tree-like structures, which start off with a trunk and then many branches, though useful for our discussions, may not have enough options to represent the complexity of the multiple arrows of time that seem to be at work in a quantum simulated multiverse.

Note that we are also concerned with higher-level descriptions of timelines, unlike the descriptions that occur at the level of particles that quantum physicists obsess about, so we need a higher-level structure that can not only represent subatomic particles, but also higher-level events in the world – i.e., a timeline.

What Is a Timeline?

You would think we'd start talking about timelines with a simple definition of what time is. As we saw in *Chapter 7: The Nature of the Past, Present, and Future*, this is a much more complicated question than it appears at first and may not be fully answerable.

In this chapter, I'd like to look at what time is in a digital multiverse, which might be an easier question to answer. But we'll start with an even higher-level, but equally important, question: What is a timeline, exactly?

Readers may be surprised that in a book about multiple timelines, we haven't attempted to answered this question yet. The truth is that I've purposely put it off because it seems like something that we all think we know. If we look it up, two definitions in the dictionary seem to give pieces of the puzzle we are looking to assemble:[135]

Timeline Definition #1:

A graphic representation of the passage of time as a line

Example: "His book, which includes political maps, timelines, and a running glossary, is the preeminent introduction to the subject."

Timeline Definition #2:

A chronological arrangement of events in the order of their occurrence

Example: "The CIA's timeline of his whereabouts has him arriving in Miami on May 28, 2001."

To combine both of these definitions, a *timeline*, for our purposes, will be a graphical representation of an arrangement of events. Note that I used arrangement and not sequence, because sequence implies an order in time, and though in our timelines we will be implying a certain order or sequence, when we're concerned with spinning off timelines and then merging or pruning those timelines, the sequence of things might get a little bit murky in a "timey-wimey" manner.

But what are these events exactly that we are arranging in a timeline? You'll realize that this is a question not unlike the one that drove Mandelbrot to notice self-similarity at different levels and fractals: How long is the coastline of Britain? The answer was that it depends on what scale you look at.

Given that there are many levels and types of events (ranging from quantum events to common personal events like marriages to big-world events like a world war, involving millions of people), how are we to know whether they should be included in a particular timeline?

The answer is that it depends on *context*. In common usage, a timeline usually refers to a context, which defines the level of abstraction of the events, which are then arranged based on this context.

A timeline of history of the United States, for example, might show only the presidents elected and the major wars the country has been in. The relevant intervals in that timeline might be years or even decades, whereas a timeline of a particular battle in World War II might be measured in days or hours. And the timeline of a criminal case might show the events relevant to the court case with intervals of minutes. Finally, subatomic events might deal with intervals that are very small, much smaller than a second.

Is there a way to represent all or some of these levels at any scale?

Like a River: Making Time Spatialized

Timelines, by their definition, are graphical representations of something that we don't experience in a graphical way. Since our intuition usually relies on spatial positioning for understanding (moving from left to right or top to bottom), we will revisit the spatialization of time, which is a technique we've seen multiple times before.

In any kind of graphical representation, particularly in a two-dimensional one, it's important to know not just what the axes stand for, but also what the intervals are and how we move along each axis. These are particulars that are often skimmed over when physicists make space-time diagrams, because it requires perhaps at least some assumptions about what time and space actually are.

Let's go back to the simple space-time diagram of Minkowski and Einstein, which is built on top of a simple two-dimensional graph, as shown in Figure 43. In Minkowski diagrams, the time axis is vertical and shows how a particle is moving with respect to the speed of light.[136]

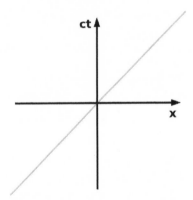

Figure 43: The format of a Minkowski space-time diagram.[137]

In these diagrams, the world always moved forward (i.e., up the ct axis), and depending on how fast a particle was traveling, its perception of time (or in Einsteinian terms, its *inertial reference* point) would change. A particle's worldline was a time-like line that showed its progression through time and space.

What we're concerned with is a timeline that encompasses events at the macro level and multiple timelines that would encompass different versions and series of events, so the

Minkowski diagram will be of only limited use to us, moving forward. In our discussion, we're going to borrow only the most basic idea from Minkowski, that of representing time as the y axis and some kind of motion or progress along the x axis. If, in an x-y graph, we assume that the future is up, and increasing a parameter t will go into the future and decreasing t will go into the past, it seems so basic that we don't usually even include these intervals.

Before we get very specific, let's start with the very simple high-level timeline diagram that makes it easy to understand what we are representing in Figure 44.

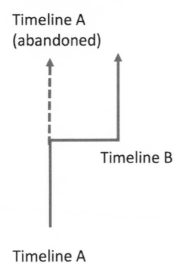

Timeline A
(abandoned)

Timeline B

Timeline A

Figure 44: An intuitive way to think about the flow of time and timelines in a digital way.

As mentioned, if the idea is that vertical travel represents movement through time, then what does horizontal travel represent? In most space time diagrams, it represents a change in position in space – a change in the (x,y,z) coordinates of an individual particle. But in a timeline graph, it *represents some event, or a change in the universe of some kind.*

But what does this mean exactly? As soon as we start to think about the axes more rigorously, we run into some challenging questions.

Finding an Origin and Intervals

Let's start with the vertical axis, which seems the simpler of the two. Clearly, if time is the vertical axis, then the intervals would be some interval of time—in fact, we think of the interval naturally as being a temporal value —from one point in time to the next.

But how do we represent this? Is it from some point of origin, like the Big Bang? Given that we don't know the exact time of the Big Bang (estimates have gone from a few billion years to our current estimate of 13.8 billion years in just a few decades), it's likely this will change again as our knowledge of the universe expands, so this doesn't seem like a practical solution.

Obviously, we need to calculate time relative to something (though we are now hinting at the fact that time may not be an absolute at all and is always *relative* to *something*). You would think that, given our understanding of the universe around us, we would have adopted a more astronomical calendar, such as a star date not unlike *Star Trek* captain's logs, which usually start out an episode with a timestamp, "Stardate 46254.7." We really haven't even agreed on a common dating system on Earth, let alone in our solar system or the galaxy (and don't even get me started on intergalactic dates!).

In human history, it has not been uncommon for religious events from the beginning of a religion to be the origin for a new calendar. For example, we use the birth of Christ as the point of origin of our most used dating system with the designations AD and BC, in combination with a solar calendar (the Gregorian calendar). Technically, this system wasn't used until the year 800 AD.

AD is derived from the Latin phrase, "*Anno Domini Nostri*

Jesu Christi," which translates to "in the year of our Lord Jesus Christ" and not, as I thought when I was a kid, After Death. BC stands for Before Christ. But a majority of the people on Earth are not Christians (though it is the largest religion by population). In a bid to make it a little less parochial and inclusive for non-Christian countries, the terms CE (Common Era) and BCE (Before Common Era), rather than AD or BC, are often used today in academic circles.

The second-largest religion on Earth by population, Islam, has a different calendar altogether, and if you have visited Islamic countries, you know that they keep track of both calendars. The Islamic calendar uses a different point of origin and a (slightly) different set of intervals. The Islamic calendar is lunar (meaning each month is exactly 30 days), and the origin is not the birth of the religion's founder, but actually an event that happened in the life of the Prophet Muhammad. This event, the flight of the fledgling religion and its followers from the holy city of Mecca to the neighboring city of Medina, was when the religion started to take off. It is called the Hijri, and the Islamic calendar designates years as AH or BH (for After Hijri or Before Hijri). The date of this in our common era calendar would be 622 AD/CE. Therefore, the year 1 AD would be 621 BH (Before Hijri in the Islamic calendar) and as I write this in December of 2020, the year, according to the Islamic calendar, is actually 1442 AH.

The way that most timeline diagrams today get around this is by not specifying a fixed point in the past but, rather, using the present moment as the origin of the coordinate system, which we might say is t=0. There is always a present moment, after all, and it gets around having to know when time began, how long it has existed, or how long it will go on. Then the area below the origin would always have *t* value that is negative (and at least nominally would represent the past), while the area above the origin would represent the future, though of course those terms are completely relative to whatever point in time we are using as the present.

Measuring Time Inside a Computer Program

Because we're talking about simulated timelines in this book, however, we can't rely on human time. Instead, we have to ask a somewhat tricky question: In a computer program, how do you move from time t=1 to time t=2?

This seems like yet another trivial question but it isn't at all, because it gets deep into the heart of the question about what time is and how digital universes work.

In the real macroscopic world, time just seems to flow—as long as the universe is on and the program is running, time will move forward. In the world of CPUs and computer programs, however, something else is driving the execution of the program: the operating system running the program, which will move the program from step n to step n+1.

With classical computer systems, the processor really only knows about its underlying clock speed and how many intervals have passed. These intervals (or cycles) are usually referred to as number of operations and measured in hertz, which means operations per second: 1 Hz is one operation (or processor cycle) per second. For example, the original IBM PC used an Intel chip that would perform at least 3.77 MHz, or 3.77 million operations per second, whereas the Intel core i5 processor, introduced in 2009, could do 4 GHz, or up to 4 billion operations per second.[138]

Technically, if you are running multiple processes (which all modern computer systems do), then each process may not necessarily know how many cycles have passed in the overall system; it may only know how operations have passed in the current *context*. Remember that context switching, which we covered in Chapter 8, means that only one process and its data are loaded into the processor at a time.

Modern operating systems will swap out programs based on memory requirements and prioritization. Let's look at a simple computer program that just runs in a loop, as shown in Figure 45 (written in an example of a high-level language like BASIC or JavaScript). Though this program has only four lines of code, it could be compiled down to a much larger number of machine instructions or operations. These operations eventually boil down to Boolean logic gates in the microprocessor.

The only function of the code in Figure 45 is to let the user input their name and then to print it out 10 times. Even such a simple program, which only lets the user input their name and prints it out 10 times in a loop, might compile down to hundreds of opcode instructions.

```
PRINT "PLEASE TYPE YOUR NAME?";
INPUT Name$;
FOR I = 1 to 10
  {PRINT "HELLO " + Name$;}
```

Figure 45: A simple computer program (in no particular language).

One set of code is usually run as a single process (or a single task or thread within a single process). In modern, multi-windowed operating systems, a process can get swapped out at any time. By swapped out, we mean that the program is put on hold while the operating system's (OS) current context switches to other programs that are running. This will probably happen initially when the program asks the user to type his or her name and waits for the user input.

This might only take the user a few seconds or it might take half an hour if he decides to go get a cup of coffee or start browsing the web. Whether it's 1 second or 30 minutes, these are gargantuan numbers when you think about billions of operations per second. So the program is put on hold by the operating system until the user types in their name. By putting it on hold, the OS will load other programs and run a number of operations of those before returning to this program. When the program gets loaded again, it will think that only one step has passed in its *current context* (though, of course, it could have consulted some external source to get the actual elapsed time; in this case, the OS should theoretically know how many cycles have passed in total across all processes).

In any case, stepping back from the details, basically in any type of simulation, the time variable *t* is usually the number of steps that have passed. These steps could be operations or some higher-level value (for example, generations of fruit flies, years, or transactions processed, etc.).

So, if we are dealing with a simulated universe or multiverse, we can think of *t* as number of *steps* that have been run, rather than any sense of absolute time that has passed.

This also helps us to deal with Einstein's special theory of relativity and time dilation – because we only know how much time has passed in the current context, without regard for how much time has passed in other contexts. The twin paradox, which we discussed in Chapter 7, becomes much more explainable in a digital world. Twin A (Adventurous Alice), who is traveling in space and is suddenly younger than twin B (Muffet who sat on her tuffet), who has stayed on Earth. Twin A, we would say, has run less steps of the program than twin B, so less time has passed.

The Horizontal Axis: The Gamestate

Let's get back to the x-axis. Unlike physicists, who are interested in the movement of specific particles from coordinates (x_1, y_1, z_1) at t_1 to (x_2, y_2, z_2) at t_2, in this book we are concerned with a graphical representation of a series of higher-level events in a timeline. With multiple timelines, we are concerned with how a decision causes one timeline to branch out into new timelines. When this happens, the whole world, and all the particles in the world, is set on another trajectory.

What is it that changes *exactly*?

In video games, we would refer to this as a change in the

gamestate (which encapsulates the entire state of the world). In fact, to create two timelines, as we saw with *SimWorld*, all we would need to change is one bit or value in the gamestate and then let the program run again.

Since the whole gamestate can be represented by a set of bits, in a simulated universe, we can think about the horizontal axis representing a change of bits in the gamestate.

In this respect, we can think of the x-axis as all the possible gamestates of the virtual world.

How could we then define distance between any two arbitrary points on the horizontal axis? In a normal integer-based axis, we would just subtract the numbers: x=50 is exactly 45 steps from x=5.

In our case, the x-axis represents sets of bits assembled into a gamestate, so the distance on the x axis would then be the number of operations that are required to change the gamestate from the first point to the second point. This means that going from, say, a gamestate of 10000000b to 11000000b is a small distance (only one bit is changed) whereas going to 11111111b is a long distance (where 8 bits have changed). This assumes a simple 8-bit gamestate like we had in *SimWorld*.

How many bits do we need to represent the entire physical universe? It would depend on the amount of information that is stored in the particles of the universe. Seth Lloyd makes the point in *Programming the Universe* that the number of bits needed for an apple is approximate to the number of atoms. Of course, with data compression techniques that have been perfected by computer scientists, the actual number might be much smaller in a simulated universe.

We could then realize that each interval on the horizontal axis represents the distance of bits between different nodes or gamestates.

Operators: Moving Across the Axis

If we define the horizontal axis by the degree of changes to each of the bits of the gamestate, we now have a great way to represent a single timeline: a timeline is a traversal of a graph of nodes of gamestates. How does this happen in information theory? Each traversal means an *operation* is performed to change the values of bits.

But what are operators in the context of a simulated reality? In a video game, gamestate changes happen in two ways:

1. Choices made by the player, or
2. Changes made by the program. These changes might happen for many different reasons, including responding to players or other procedures that are part of the game.

Now let's visit how movement across the x-axis might happen in a more formal way. Let's suppose each interval of time causes a choice to be made, which could be a single bit to change or stay the same. For a single interval of time step t, you can only navigate from one gamestate to one that is a distance of one operation away. We now have a more formal x-axis, and a way to represent different timelines on the same graph (at least theoretically).

But how do computer programs decide which bits to change? This is done via those Boolean operators that we discussed in Chapter 10. Examples are AND, OR, and NOT gates, which take inputs of one or two bits, and output a bit based on their logic. Since any computer program can be thought of as a series of gates performing Boolean logic, the natural way to think of going from time t to time $t+1$ is to have some program or sets of programs that define the transitions of the bits.

One way is to think of a fully deterministic universe where the whole universe runs via a fixed set of computer programs processing the bits of the gamestate, not unlike a cellular automaton.

This means we can't just jump from any series of zeros and ones to any other set of zeros and ones; we can only jump to those allowed by the operators that let us navigate on the x-axis. In a practical sense, this might mean that the bits of an apple A shouldn't change to the bits of a pear B in one instant (i.e., one step of time), though the object in the hand of your avatar inside a video game might change to a pear after a million operations at the level of the CPU.

Another way to think about it is in a multiplayer video game where the players have free will and are able to make choices for their characters (referring back to the discussion in *Chapter 9: Simulation, Automata, and Chaos* about free will and randomness in cellular automata).

Whether we allow for determinism or rely on randomness or allow an external agent like a player (i.e., a conscious entity) to initiate a change of bits, the end result is the same: we are still watching bits change from one gamestate to the next.

In its essence, that is the process of computation: transforming bits using operators.

Big Events in Timelines

Getting away from bits now, though, when we want to branch timelines we think of two timelines as having different series of events. For example, Nelson Mandela might end up dying in prison in the '80s in one timeline; in another timeline Mandela gets out of prison and becomes the President of South Africa in the '90s. In one timeline, JFK is assassinated, and in another timeline, he isn't. In one timeline, the Allies win the second world war, and in another timeline, the Axis powers win the war.

These are of course, big events. If we were to represent the whole world digitally, it would take more than one-bit changes between these timelines. It would also take more than one step of computation. In fact, we could define an *event* as a series of bitwise computations on an underlying gamestate. The result of the event would be moving to a new gamestate. This could be one bitwise change (a quantum event at a subatomic particle level) or a change of many bits in a combination of operators.

So, we can now trace the route of a timeline in our 2D graph, which has a vertical axis for time and a horizontal axis for gamestate, both of which can be defined in terms of computation. An operation that changes one or more bits moves us along the horizontal axis and up the vertical axis. How far up? This depends on how much time the event takes. And that depends on how many bitwise operations we need to get from one gamestate to the next.

In a way, this idea that both the vertical axis and the horizontal axis of our timeline graph consist of running operators on bits is a simplification and a complication. In a sense, there is almost no difference between the axes—the time needed to move from one gamestate to another is just the number of operations needed to get from one gamestate to another. Time and space are both doing the same thing: computing bits.

This brings us to the realization that perhaps a simple 2D graph isn't the best way to represent the possible gamestates in a multiverse, since both axes are doing the same thing. In this context, it's almost more appropriate to think of a network of nodes that are being visited by each program. Let's look at an alternate way to represent this.

Branching and Merging: The Way Is the Multiway

Stephen Wolfram, one of the pioneers of cellular automata research (whom we mentioned in Chapter 9: *Simulation, Automata, and Chaos*), has been a big proponent of the idea that the physical universe may actually be a computational universe consisting of smaller programs.

In his book, *A New Kind of Science*, published in 2002, he attempted to lay the foundation for how rules (as defined by elementary cellular automata) and computation might account for a different way to think of the science of the physical world. This new way was based on computation rather than on the prevailing view that mathematical equations were the way to predict the physical universe.

As a follow-up, in 2020 Wolfram launched the Wolfram Physics Project to show how the laws of physics, such as relativity and quantum mechanics, might actually be derived from simple computation rules like those in cellular automata.

In his more recent work on which the Wolfram Physics Project is based, *A Project to Find the Fundamental Theory of Physics,* Wolfram introduces the idea of branchial space and multiway graphs, tools that perhaps might give us a more complex yet realistic way to think about multiple timelines as a network in which gamestates can not only branch out, but merge again as well.

In Wolfram's multiway graphs, he uses strings like ABCD to represent different nodes in space (and, as we shall see, in time). These letters could, however, corresponds more or less to what we have been calling the gamestate, which we have been representing as bits of information.

If we ignore the human element of consciousness for now (which many physicists try to do) and consider the universe as a set of quantum choices, then you can think of a multiway graph as all the possible configurations of the particles in this particular digital universe. You can then apply a rule to go from one node to another, or what we have been calling operators.

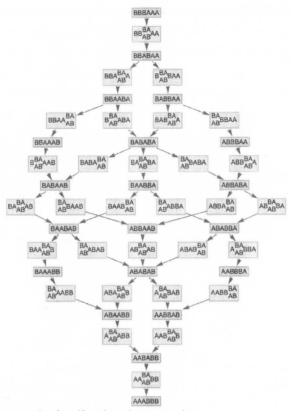

Figure 46: An example of Wolfram's multiway graphs.

The definition of the edges in this network of nodes constrains what is allowed to happen. In Wolfram's graphs, there is only a single rule, as in his elementary CAs, but this may or may not be the case in the physical universe. The important idea is that the multiway graphs he defines are a perfect way for us to think about the multiverse and how to diagram it to create a Multiverse Graph.

As shown in Figure 46, a multiway graph is a type of *hypergraph* that consists of series of nodes and edges. A hypergraph is simply a general term for a network of nodes and edges. In a hypergraph, each node can connect to many nodes, depending on the rules that constrain motion across the nodes.

To simplify, let's zoom in on what might be happening in a

single node (or single gamestate).

In these graphs, even using simple rules, you'll see that it's possible to BRANCH and to MERGE.

In Figure 47, we see how a gamestate with four digits (using Wolfram's convention of letters rather than bits) could branch into four other values. Each new node (or branch) has just one changed digit. (In our case, the digits are A and B, but an actual multiway graph can use any combination of digits.) In the end, these will be represented as bits.

If we then assume we keep going in time (in this case, time is going downward, following Wolfram's convention and the convention of cellular automata), then we see that different worlds can also merge, because you can end up back at a previous value, or two different gamestates can, via two different operations, end up with exactly the same gamestate.

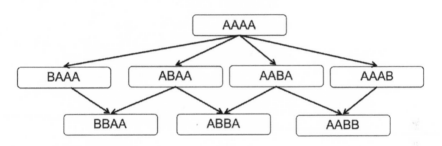

Figure 47: A multiway graph that shows branching and merging over time by changing one bit's value.

For our purposes, in a Multiverse Graph, we are going to simplify to say that each node is a gamestate of the world (i.e., all of the bits in the entire universe). While this isn't exactly how Wolfram defined his which way graphs [139], it gives us the basis to talk about multiple timelines and the multiverse.

This also means that each of our nodes carry quite a bit of information, but these are theoretical nodes and not necessarily physical ones. Since the possible configurations (or gamestates) are limited by the number of bits (or digits), this must be a finite graph. [140]

> **Multiverse Graph:** A graph of all the possible node
> values of the gamestate of the universe.

Defining a Timeline More Formally

Travel between nodes can represent a quantum choice and a change of some individual item – like a bit, a particle, a digit. Similarly, any set of nodes and edges could be combined to encapsulate higher-level definition of an "event" – any event really.

This simple type of diagram demonstrates an aspect of multiverse theory that may be surprising: universes can not only branch, they can also merge together, which means there could be multiple possible paths to get from one node to another.

This also leads us to a computational way to express timelines and the multiverse.

> **Timeline:** In a network of all the possible states of the
> world (the Multiverse Graph), a timeline is a subset of
> nodes and edges that are connected to each other in a
> specific way. The time that passes is the number of edges
> that need to be traversed, or computed.

At first glance, this might seem like a lot of work for defining something we already defined at the beginning of this chapter and that everyone thinks they know intuitively. But if you look closer, our original formulation of a timeline was great for intuition but difficult to pin down; the axes were fuzzy, as were the intervals on each axis.

We now have a way to define a time interval, which is a change of state of the bits of the world. This could be limited to changing one bit at a time or be a simultaneous update of some number of bits in a higher-level operation (which could be compiled down to smaller one-bit changes in the same way that computer programs are compiled down to more primitive operations). The more bits that can be changed in a single time, the better they define the parallelism of the computer system that is running the multiverse.

The amount of time that has passed on the timeline is the number of edges that have passed. These are analogous to steps of computation—and could correspond to any amount of time in world (i.e., seconds, days, years, nanoseconds, etc.) as long as they are some multiple of the underlying computational step, or clock speed, of the processor.

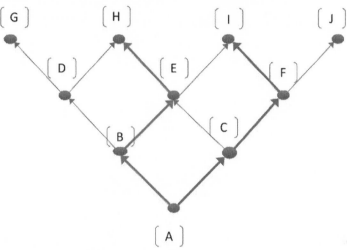

Figure 48: Moving from A to B or A to F?

Some Observations About Timelines in the Multiverse

We now have a way to think about multiple timelines that matches our earlier intuition of worlds branching and merging and progression along these timelines.

This is particularly relevant in a digital universe or a digital multiverse. Note that this graph is not of a single universe but of the whole multiverse that is possible.

In Figure 48, for convenience we have simply labeled each node with a letter—A, B, C, D, E, F, G, H, I, J—though each node is really a possible gamestate (or bitstring of all the particles in the universe). We can now make the following observations:

- The present could be anywhere on the multiverse graph that we choose to focus, from which we are measuring.
- The past then is any nodes we have passed through.
- The possible pasts are any feasible path that brings us to the current node from some arbitrary previous node.
- The future then is any nodes we decide to travel to from the current node.
- The possible futures would be the subset of the graph of nodes that are reachable by moving up the edges from the current point.

In Figure 48, going from {A, C, F, I} would be a legitimate timeline on this graph to go from A to I. Similarly, going from {A, B, E, I} would also be a legitimate way to get from A to I, though following a different path of intermediary nodes.

We now have, in time, something akin to the associative property of addition or multiplication that we learned in school: you can add or multiply numbers in any order and still get the same result—for example, $(3+2)+5+4 = 3 + (2+5) + 4$. They both add up to 14 in the end, though we took different computational paths to get there.

Although easy to understand for simple addition, it brings us to a surprising revelation (at least for those who haven't studied the delayed-choice experiment): If we are in the world with a particular arrangement of particles at point I, there are multiple possible pasts! You can't say which past is a legitimate past, because they both are. In the example that I just used, from point I, points E and F are both legitimate past points, but not on the same timeline. We now have a computational model to define the Mandela effect and the delayed-choice experiment.

Simplifying the Multiverse Graph for Ordinary Timelines

Now that we have a more formal way to think about and represent timelines on multiverse graphs, we can still go back to our simpler timeline diagrams, as shown in Figure 49.

Zooming in on the two timelines, we would see they are actually a connected set of nodes, and the node where the two paths connect is the merger point, or where the two paths diverge is the branching point.

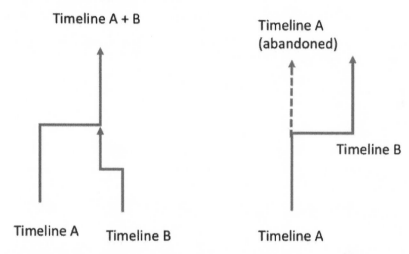

Figure 49: Simple timeline diagrams.

We now have a more formal way to think about the multiverse in terms of computation. Yet if we were simply running a single CPU and only traversing the nodes one at a time, there wouldn't be a need or implication of a multiverse—no multiple timelines, no saving and loading of gamestates, no probabilities of multiple futures or multiple pasts, and no Mandela effect.

All of that wonderful weirdness and complexity of quantum mechanics dealing with time and collapse of the probability wave and the quantum multiverse now only comes out because of the way we traverse the Multiverse Graph: our Core Loop, which we will turn our attention to next.

LOKI AND THE MARVEL MULTIVERSE

In 2021, as I was writing this book, Marvel Studios (a subsidiary of Disney) rolled out a new streaming series based on their comic book worlds that introduced a new multiverse theme. The show was named after Loki, the God of Mischief, and the adopted brother of Asgardian God of Thunder, Thor.

In the first episode of the first season, Loki (Tom Hiddleston) finds himself in a strange place: the TVA (the time variance authority). It turns out that he is a *variant* (or a particular type of what I like to call a time instance) and that his actions have branched a new timeline. The TVA was set up to "preserve" the sacred timeline, which means that anytime a new timeline branches off that is not part of the main timeline, they prune the timeline by bringing the variant to the TVA. Owen Wilson, who plays Mobius, is one of the agents, tasked with tracking down variants. At the TVA, there is a monitor which shows the timeline evolving and whenever there is a new unexpected branch, the TVA is alerted to the variation on its screen, and its agents travel through time to prune that timeline.

There is a particularly troublesome variant—in this case a female version of Loki that likes to call herself Sylvie (Sophie Di Martino). Sylvie travels through time, hiding from the agents of the TVA and causing all kinds of mischief in the timeline. In an interesting twist, Loki figures out her preferred hiding spots in time: just before major apocalypses, where her presence doesn't cause an unexpected timeline branch.

It seems that the TVA is set up to prevent the type of multiverses that we have been discussing in this book. However, eventually, Loki and Sylvie team up with other variants to try to find out what's really going on behind the TVA.

They eventually discover "He who remains" (Jonathan Majors), who set up the TVA in the first place to preserve a single timeline. It turns out he was a scientist who discovered the multiverse, and that there were many versions of him doing the same thing. Eventually this led to a multiversal war, and to avoid that, the character (who is called Kang the conqueror in some of the Marvel comics) has set up the TVA to preserve a single timeline and to make sure that other timelines aren't branched off, effectively pruning the multiverse before it gets started.

Most interesting from our point of view, we see the multiverse emerging in a graphical format on a screen every time there is a variation. It is a beautiful tree-like structure shows multiple timelines and branches developing in real time.

Chapter 12

The Core Loop as Search

"By our choices, we each thread our own separate way through the maze of possible worlds, bypassing equally real alternatives with equally real versions of ourselves and others, selecting the world we must then live in."[141]

—Hans Moravec, *Simulation, Consciousness, Existence*

In this chapter, I'd like to delve into the mechanics of the Core Loop that I have mentioned several times thus far without giving a more formal definition. This will sum up the parts of this book that have been focused on individual pieces of the overall picture, and we'll switch to the big picture in the last part of this book.

Running Simple Simulations

Let's suppose we were simulating an ant colony. This would be a very simple, nongraphic simulation using a simple equation that is applied again and again to determine what the population of the colony would be after a number of steps. Depending on what type of colony we were simulating (ants, fruit flies, etc.), the interval of generations could be a year or a season or a day. The key variables would be the initial population and growth rate (including birth rate and death rate).

Of course, real-world populations are rarely this simple; there are always unexpected ebbs and flows. We could add some complexity to our simple model by having the growth rate (birth to death) vary based on the size of the population: if it gets too large, there won't be enough food, so the death rate goes up, for example.[142]

Another way to accomplish the same thing would be to add some rules, as we do in computer programs. In simple computer programs, a series of IF-THENs could be used to moderate population growth. Using this kind of program, after some oscillations, the population number is more likely to settle down to a range of equilibrium.

But in this simple deterministic simulation approach, there is no multiverse to speak of, per se. If we wanted to try out different variables, or perhaps add an element of randomness, we could come up with different results based on each set of variables. We would have a probabilistic set of possible futures.

We could go even further by allowing for this change of variables after each step (or generation). In this case, rather than, say, just five results for five growth rates (given an initial population), we would have five timelines spawning off after each step, resulting in a much bigger tree of possibilities that starts to look more and more like the Multiverse Graph of the previous chapter.

Why would we do this (i.e., add random variables and try to run separate timelines)? One reason is to figure out what the probabilities are of different outcomes. In what's called a Monte Carlo simulation, you run the same process many times to see what the outcomes are, and then you plot the results as a distribution. If the numbers cluster in particular locations, you can confidently say these locations are more likely outcomes.

The other reason simulations are run is to find the optimal outcome. As we mentioned in Chapter 10, one of the areas that quantum computers are proving very useful for is the simulation of molecules for drug discovery. These simulations are of exponential complexity. It would take classic algorithms many years to run all the possible combinations to find the one that is most optimal.

Defining the Core Loop

If we really wanted to find the optimal outcome for some simulated entity at each step of a Multiverse Graph, we would have the basic structure of the Core Loop. At each point, we would try out different values for next steps either deterministically or randomly or through some element of choice by people outside the simulation (players or simulators). By doing this, we are effectively spawning off different timelines from a current point in time and then figuring out the best path to take. Once we have decided which branch to follow, we might repeat the same process, spawning off different timelines and figuring out which one was optimal as a next step.

You could define the Core Loop (as a classical algorithm) as optimizing the future with the following steps, assuming a simple classical gamestate:

- SAVE CURRENT GAMESTATE A
- CREATE_BRANCH B
    - ALTER SOME VARIABLE in GAMESTATE A – and call this GAMESTATE B
    - RUN PROGRAM starting with GAMESTATE B for n number of steps
    - At step n, calculate the score of this path (i.e., BRANCH B)

You could then rerun this process by recording the result of BRANCH B and then reloading A, changing the gamestate, creating BRANCH C, and so on, up to the maximum breadth and depth of possibilities that could branch out from A.

Eventually, we would record the score of each branch and then make a choice. Note that having a limit of branches we could explore and a maximum depth (or number of steps to run each branch) is important; otherwise, we end up with the infinite-loop problem: the program would never come back with a result—each branch would run forever.

An Example of a Video Game Algorithm: Minimax

This kind of process—of trying out different values and evaluating the best path to follow and then repeating the procedure at each step—is a familiar one, and we use it in computer games and other simulations all the time. In computer science, we refer to this as traversing the graph, or searching the graph for an optimal outcome.

How would we figure out which branch was optimal? We would need some *fitness function* that assigned a desirability value (the score) to each branch. This fitness function would be a function that evaluates the relative strength (or weakness) of each possible path or timeline by giving it some relative value, usually expressed as a number.

The nature of the fitness function varies with the games we are playing. For example, with the game of checkers, the fitness function would be very simple: the number of pieces that remain for you and for your opponent determine how optimal a move would be. In chess, the fitness function wouldn't just be based on how many pieces you have but which pieces you have left (for both you and for the opponent), given the relative strength of those pieces. If you lose a queen, for example, by going down one path, then that path is not a good one.

I was first exposed to this in building games like chess and checkers with a simple fitness function in what's called the Minimax algorithm, an example of which is shown in Figure 50. It's called that because from the current state of the game, we evaluate a fitness level of the board at each step for the player and the opponent. The goal is to minimize your opponent's fitness function score and maximize yours.

In the previous chapter, we showed how our simplified idea of timelines could be more formally expressed and understood as a Multiverse Graph of nodes. If we think of each of the nodes as a gamestate, the network shows us the list of possible futures and possible pasts from every point in time. In a game, the movement is accomplished by choosing a move, which would be represented as a change in gamestate.

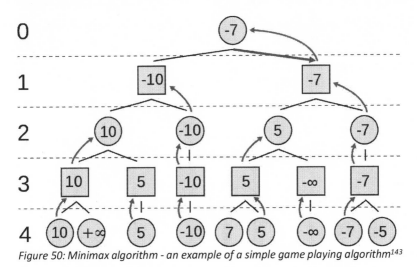

Figure 50: Minimax algorithm - an example of a simple game playing algorithm[143]

Traversing the Graph: Breadth First versus Depth First

The Core Loop lets us maximize the possible future by examining all of those futures. In the case of our Multiverse Graph, the nodes are gamestates representing the world, and the possible moves are changing the bits by using operators. In a simulated universe, this would mean actually traversing the graph and accumulating some sense of the desirability by letting the simulation run for a number of steps.

There are many different ways to traverse an arbitrary graph, but there are two basic algorithms that are simplest and most often used. They basically define the order in which we might go through the possible nodes we can reach, called breadth-first and depth-first.

The simplest of these is to use a breadth-first algorithm, as shown in Figure 51, which shows the order in which you would search the nodes of the graph, starting at 1 (in the first row) and moving to the second row (nodes 2, 3, 4) and then the next row (nodes 5, 6, 7, 8, etc.). The problem with this approach is that although this search is manageable for a simple graph, it can get staggeringly complex as the number of connecting nodes increases. If you have too many nodes to search at each level, this effectively makes it very difficult to search to any depth in any reasonable amount of time.

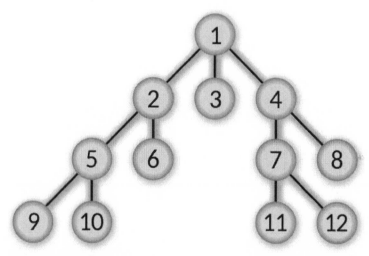

Figure 51: A breadth-first search of the graph. The numbers represent the order in which nodes are evaluated.

Furthermore, it's unclear how we can evaluate the fitness of each of the nodes at each level without going further down each individual path. For example, to see how promising the first path starting with node 4 might be, we have to reach nodes 7, 11, and 12, which won't be reached for a while when using breadth-first. In a more concrete example, sacrificing a queen in chess might give us a very low score for the board in the next move, but if it leads you to checkmate your opponent in three moves, then that path may be the best one.

The other algorithm, depth-first search, may be better when the fitness of a particular node depends on what might happen down the road.

A depth-first search order is shown in Figure 52, where you can see that we traverse the nodes by going deep down one or two paths, to some predetermined maximum depth, before backing up to go down the next path. In this case, the depth is 3 (because there are three levels below the starting node that are searched). One timeline is explored all the way to the depth level of steps in the future.

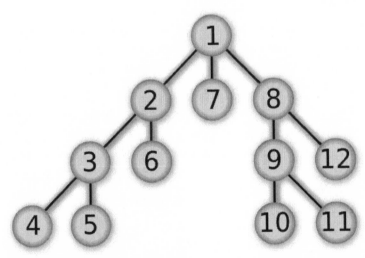

Figure 52: Example of depth-first search.

Why would you want to stop with a certain maximum depth? Because the deeper you search, the more computing resources will be taken up and the slower the whole thing will be. In an open-ended multiverse, it might seem that the graph has an infinite depth, so the computation would literally never stop. So the maximum depth is the boundary condition we use, but at each node we are exploring, we apply exactly the same algorithm.

Recursion and a Depth-First Search

In fact, if our Core Loop is what's actually happening, and multiple universes are branching out, this works very similarly to recursive algorithms. The thing about recursion is that it can be used to generate very sophisticated results. We explored fractals in Chapter 9. The way to generate fractals, which are about self-similarity at every level, is to use recursive algorithms at each level.

Recursion is a way for a computer program to run the same version of code on different variables, but for each run to be unaware of the others. For all effective purposes, each run has its own context and is mostly ignorant of the other runs, except that it gets some values passed in, which it processes and passes back the answer to the previous version.

Recursion is ideally suited for depth-first searches because it goes deeper into the graph and calculates the next value, returning the value of each branch to the starting point. Recursion is a way to simplify a big problem to a slightly smaller version of itself, calculate the answer to that smaller problem, and use that answer to calculate the final answer.

Let's use a simple example that calculates the value of an integer raised to a certain power (an exponent) using recursion. Suppose you wanted to calculate 2 raised to the 8th power. The formula is quite simple for any value of n as the exponent:

$$2^n = 2 * 2^{n-1}$$

So, 2^8 is actually 2 multiplied by 2^7. If we can call a function to calculate that smaller problem, we can simply multiply it by 2 and get the answer we want. Similarly, $2^7 = 2 * 2^6$, and so on, until we get to $2^1 = 2 * 2^0$, where 2^0 is defined simply as 1. This is the boundary condition and the point at which the code stops applying the same algorithm and returns the values up to the previous version of the algorithm.

If you've never seen recursion before and are learning to program for the first time, it can seem a bit strange. It can also run in an infinite loop unless you set the right boundary condition. Like the smallest doll in nested Russian dolls, there must be a smallest level.

Pruning the Tree: The Genetic Algorithm at Work

You may have noticed that the process I've described here, in computer science terms, of building and pruning a tree of possibilities is very similar to a process that we've all heard of: evolution. Evolution, though starting off as a theory of biology, has made its way into information science as a method that can be used to optimize really anything.

Evolutionary theory is now used in many fields, not just in the evolution of organic entities. The idea of tree-like structures branching off based on changes in variables, with the best branches surviving, is almost universal whether we are talking about physical entities or abstract or concrete intellectual ideas. Religions branch out from other religions. You can map that like the tributaries of a great river, and some of those branches die out while others become bigger. Languages also branch out. Tree-like structures seem to be fundamental in nature in the physical three-dimensional physical world. They are also essential in the intellectual world of ideas.

What the Multiverse Graph and Core Loop are implying is that tree-like structures also provide an essential view of space-time. The tree branches project forward in time until that branch of the tree is pruned or becomes the main branch.

Daniel Dennett, in *Darwin's Dangerous Idea*, wrote, "Little did I realize that in a few years I would encounter an idea—Darwin's idea—bearing an unmistakable likeness to the universe's acid: it cuts through just about every traditional concept, and leaves in its wake a revolutionized world-view, with most of the old landmarks still recognizable, but transformed in fundamental ways."[144]

Richard Dawkins, of Oxford, called this idea, that evolution could be applied in nonbiological settings, universal Darwinism. In computer science, we call it an evolutionary or genetic algorithm.

First coined by John Holland at MIT in the 1940s, a genetic algorithm is an interesting mix of biological techniques and information science that shows the interrelatedness of these areas. The genetic algorithm, which is one of a class of computer science algorithms referred to as evolutionary algorithms, is a way to search a space efficiently to find optimal solutions. The operators are borrowed from biology, such as mutation, crossover, and selection.

Let me recall a layman's description of Darwin's idea of natural selection, which gave nature a mechanism to find the best or most fit version of a species through experimentation. Genetic mutations caused small variations between members of a species, and then reproduction produced new combinations of these genes, and the process continued. Over time, those that were the most fit would survive, their lines going on, whereas those that weren't, would effectively be branches of the tree of life that were pruned.

Genetic algorithms are "commonly used to generate high-quality solutions to optimization and search problems."[145] The idea is to search a candidate population of entities to find the ones that are likely to be the most fit for a task. Starting out with some initial version of the entity, changes are made to the entity, and each change is evaluated against a fitness function. In some cases, the most fit are then cross-bred (merged) to see which of the offspring might be more fit.

To implement a search space effectively as a genetic algorithm, you need at least the following:

1.   Digital representation of the state. A gene might be represented as a series of digits in biology. It turns out

that most problems can be represented as strings of bits, just like our gamestate.

2. A fitness function. You would need some way to evaluate each step of the future.

That genes are easy to think of as information is no coincidence. Genes were theorized as the smallest units of information that were used for heredity well before DNA was discovered. Although the mechanics of crossover are borrowed from biology (combining bits in chromosomes), the analogy with our Multiverse Graph is pretty clear. Each node in the multiverse is represented as a gamestate, which can ultimately be thought of as a series of bits representing the state of the universe.

What happens is that as each gamestate branches out, certain bits change in the gamestate. There may be many possible changes that can be made in one time step, depending on how we define the time interval and how many operations are needed.

This process, then, is repeated and each result is then evaluated using the fitness function. We see that in nature, searching for the best solution seems to be a process that is ongoing in the ever-evolving tree of life, and the Core Loop would imply this same process is at work in the evolution of the universe itself.

Revisiting the Core Loop

Our Core Loop is kind of like a sophisticated genetic algorithm, with some randomness specifying which paths we should explore and which ones should be abandoned. This is accomplished with a relatively well-understood mechanism: recursion that is used up to a certain depth to try to find good outcomes by repeating the same process downriver from where we are. The process works the same no matter where you are in the network graph. We can use this to implement our core loop.

The Core Loop that results in multiple timelines, then, is akin to a recursive search of the multiverse graph that looks for outcomes by following timelines to a certain depth and then backs up to the present. It then prunes those paths that don't seem optimal as it starts the next level of the search, not unlike a genetic algorithm.

Of course, as we saw in Chapter 10, these possibilities can be explored in parallel rather than serially, which allows for almost instantaneous evaluation of a very large number of possible alternatives, because the computation is being done in parallel worlds. In fact, that is what quantum algorithms seem to be doing.

The quantum Core Loop is a recursive process that can be run in parallel based on different futures from the current point in time, and all of these futures are evaluated very quickly to find the ideal timelines to move down next as we traverse the graph.

In the next chapter, we'll plunge into what some other physicists have said about this process and provide some interpretations, but for now, we have a mechanism: The universe is a recursive loop looking for better outcomes.

BLACK MIRROR—HANG THE DJ

In an episode of the British sci fi series, *Black Mirror*, two youngsters looking for love, Frank and Amy, are matched by an automated dating app called Coach. Coach not only matches couples, it prescribes how much time they are to spend with the other person, which could range from a one night sexual encounter to living together for months or years. The goal is for the AI, Coach, to find a suitable lifetime partner on pairing day, and the system supposedly has a success rate of 99.8%.

Amy and Frank are matched for only 12 hours and are then matched with others the next day. Both are matched for longer periods of time with other people, Frank for a year and Amy for nine months. They meet each other again at an event where another couple talks about a successful lifelong pairing, and both seem unhappy with the partners that Coach has given them for the next few months. After Amy's nine-month matching ends, she ends up in a succession of short pairings that she finds unfulfilling. When Frank becomes available again, they are matched and agree to enjoy each other's company and not worry about their matched time together. When they are both close to their pairing day, Amy convinces Frank the night before to run away together rather than waiting for Coach to pair them with other people.

During this conversation, Frank and Amy discuss how odd it is that they don't have any memories before these relationship matches. Eventually, they realize that something strange is going on with the AI Coach and the world they are in.

They rebel and try to escape.

When they do, they realize that they were in a simulation that had been run multiple times. In the real world, they are actually meeting each other in the real world in a bar with a DJ, and the simulations are being run 1,000 times to see what they would do. They find out that in 998 of the 1,000 simulations that have been run, they decided to run away together rather than wait for the AI to pair them with someone else: a pretty good sign that they are compatible with each other.

Part V

The Big Picture

I am sure, as you hear me say this, you do not really believe me, or even believe that I believe it myself. But nevertheless it is true.

        – ***Philip K. Dick,*** *Metz Speech (1977)*

Chapter 13

The Upshot—The Universe Evolves Through Multiple Simulations

Time is forever dividing itself toward innumerable futures.
*– **Jorge Luis Borges**, The Garden of Forking*
Paths[146]

In this chapter, we try to bring together the various threads we have been pursuing related to the reasons why the universe might be spinning off multiple worlds and mechanisms of how it might work. Those reasons and mechanisms have everything to do with computation.

Starting and stopping timelines—in essence, the Core Loop process I describe in this book—may be a natural part of how the universe works as it computes. As a result, the universe is creating patterns in time as well as space. These patterns in time aren't typically visible to those of us caught inside a particular timeline, but would be clear if we were observing the universe like we observe a computer simulation from outside the sim.

The baffling findings of quantum mechanics naturally lead us to speculate about a simulated world. And we know that a quantum computer is capable of running many timelines at once—just as the universe seems to do. Moreover, the Core Loop, a conjecture we have been exploring in some detail, lets us interpret quantum mechanics and our universe in a different way:

Rather than simply creating infinite parallel universes every time a quantum measurement is made, these different universes are in fact part of a computing process that relies on quantum parallelism to compute some set of results, branching off timelines and merging them when the process has figured out the *best* outcome.

Although this might seem like a highly speculative proposition, it turns out that some physicists that have been looking at the mysteries of quantum mechanics and have come to similar conclusions. Of course, this conclusion—the idea that we live in a simulated multiverse—begs the question: What would be the purpose of a simulating multiple worlds?

As discussed in the previous chapter on the Core Loop, simulations are run multiple times to come up with the most likely and/or most optimal outcomes. This is particularly true for computationally irreducible processes in which the only way to know these possible outcomes is to run the process to a certain point and observe what happens each time.

In our case, that computationally irreducible process is no less than the evolution of the universe itself. And to echo Voltaire (who said that it would be no more surprising that we live many times than it is that we live once), would it be any more surprising to live in multiple simulated timelines (the simulated multiverse) than it would be to find that we live in a single simulated timeline or universe?

In fact, the mysteries of quantum physics may show us that this is a better explanation of the facts than either a single simulation running continuously or the more traditional materialist hypothesis: that the universe is physical, that it runs forward in one direction only from the past to the future, and there are no such things as multiple pasts or futures.

Finally, the idea of a simulated multiverse provides a surprising explanation for the Mandela effect. In this surprising explanation, people are remembering slightly different timelines, based on slight variations in initial conditions, which have then been merged to arrive at our current gamestate.

Why would they be merged? As we saw in countless examples, the possible nodes of the multiverse are really different gamestates. Call them different arrangements of particles or bits, it doesn't matter since, as Wheeler and Deutsch reminded us, it's all bits (or qubits). Each arrangement is a node in our Multiverse Graphs, an idea that is meant to illustrate visually the different paths that different universes might be taking.

In this model, the Core Loop is a computational process that is inherent in how the universe works. The universe spawns multiple timelines as multiple processes that are each exploring slightly different paths through this very large (though finite) set of nodes.

These multiple histories and futures are then consolidated into a single branch that we observe as our particular reality at this time. This doesn't mean that the other branches don't exist— they were as real as this branch is, at least while it is being run. What we think of as "now" and "the past" and "our future" are really just branches of a large tree like structure in time. In fact, we ourselves may be time instances on a branch that is currently being explored in a depth-first search, in which case we might be destined for a merging or pruning in our future, unbeknownst to this specific instance of us.

"The Garden of Forking Paths"

A short story written in 1941 by Argentine writer and poet Jorge Luis Borges, titled "The Garden of Forking Paths" ("El jardín de senderos que se bifurcan"), seemed to foreshadow the idea of multiple universes as paths through time.[147] The story has inspired generations of physicists and science fiction writers to use this metaphor when trying to explain the quantum multiverse idea, and I bring it up here as a colorful way to visualize branching timelines.[148]

In this story, we learn of events through a statement from the main character, Dr. Yu Tsun, who is living in London as a spy for the Germans during World War I. Dr. Tsun, it turns out, is a direct descendent of Ts'ui Pen, a former governor of Yunnan province in China. Ts'ui Pen became famous because he gave up his provincial governorship and spent 13 years working on a novel and constructing a labyrinth "in which all men would lose their way."

Upon Ts'ui Pen's premature death (a murder, no less!), he left only a mysterious, jumbled manuscript that was published by a monk who was executor of his estate. He left no signs of an actual labyrinth but did leave a mysterious letter suggesting: "I leave to various futures, but not to all, my garden of forking paths."

During the war, Dr. Tsun goes to meet Dr. Albert, an elderly sinologist, outside London on an errand related to his spy work for the Germans.[149] Oddly enough, as soon as he enters the other man's house, Dr. Albert asks him if he is there to see the garden, which he calls the garden of forking paths. This jogs Dr. Tsun's memory of his famous ancestor's unfinished book, of which Dr. Albert had received a copy from Oxford and had been studying.

At first, Dr. Tsun dismisses Albert's study of his great-grandfather's book as fruitless: "The book is a shapeless mass of contradictory rough drafts. I examined it once upon a time: the hero dies in the third chapter, while in the fourth he is alive."

At this point, Dr. Albert triumphantly informs Dr. Tsun that he has, in fact, solved the mystery of his famous ancestor and his labyrinth. *The Garden of Forking Paths* was not only the book that Ts'ui Pen was writing but it was also the labyrinth that Ts'ui Pen was meticulously constructing.

Rather than having a character choose one alternative, abandoning other possibilities, which is what normally happens in novels, in *The Garden of Forking Paths*, every possible alternative was explored. All of the possible outcomes of an action occur simultaneously, creating forking paths. Albert has also figured out, just as we have been stating in the past few chapters, that sometimes alternative paths converge, meaning that they had a different "chain of preceding events."

In essence, the twists and turns of Ts'ui Pen's labyrinths were lines in an invisible, complex pattern across time. Dr. Albert tells him:

> The Garden of Forking Paths is a picture, incomplete yet not false, of the universe such as Ts'ui Pen conceived it to be. Differing from Newton and Schopenhauer, your ancestor did not think of time as absolute and uniform. He believed in an infinite series of times, in a dizzily growing, ever spreading network of diverging, converging and parallel times. This web of time—the strands of which approach one another, bifurcate, intersect or ignore each other through the centuries—embraces every possibility.

Dr. Albert points out that in some of the paths, he exists but Dr. Tsun doesn't, while in others it is the opposite, and in some, they both exist but never meet. They happen to be in one of the timelines where they both meet. In one timeline, they meet as friends, and in another they meet as enemies. (I won't get into the politics of the story but Dr. Tsun is a spy for the Germans Dr. Albert, so they are technically enemies in the timeline of the story.)

In short, Ts'ui Pen seemed to be describing a universe where multiple timelines are branched off based on all possible values, very similar to the MWI of quantum mechanics. Moreover, it also showed the idea that many of these paths may not survive as independent timelines but might merge with other timelines. These are the exact ideas we have been exploring in this book (though of course without the reference to computer simulations, quantum mechanics, or science fiction).

The simulated multiverse, I would assert, is an updated

version of "The Garden of Forking Paths," the "labyrinth in which all men would lose their way."

AI, Video Games, and Self-Play: Running Multiple Simulations

If "The Garden of Forking Paths" is a good way to zoom out and visualize the big picture whose particulars we have been building up, we are still left with the question: Why would someone want to build such a simulated multiverse? Why would someone want to run similar scenarios over and over again, sometimes revisiting the same paths with different parameters? One answer would be to learn from the different choices, and that is illustrated by today's game-playing AI.

When Claude Shannon laid out the ways in which computers might eventually play video games in the 1950s, he suggested several layers of complexity and achievement as the AI became more sophisticated. One of these was a game in which we program in the rules, and the computer uses those rules to play; Shannon himself created one of the first computer systems that could play chess, building custom circuits to accomplish this task. Another layer of complexity Shannon suggested was games in which the computer could learn the rules by playing and they wouldn't have to be hand-coded. Yet for many more years, AI was trained to play chess by programming in the rules—this seemed to be sufficient to create a competitive AI chess opponent.

It was thought that a game like Go, however, would be much more challenging for an AI to tackle, because it's much harder to articulate strategies that could be used by computer code to evaluate the different moves, and the number of possibilities is truly staggering.

In 2014, Google introduced the AlphaGo software, built on its DeepMind AI platform. AlphaGo used a hybrid of programmed rules and tree-based search. In 2015, AlphaGo defeated a professional Go player for the first time. Google kept modifying the AlphaGo software. One version, used a hybrid of search algorithms, programmed rules, and self-play reinforcement learning and became, over time, even better than the original AlphaGo. It was able to defeat 18-time world champion Lee Sedol in 2016. And in 2017, AlphaGo Master surpassed this version and beat the #1 ranked player in the world, Ke Jie.

Finally, the Google engineers did away with the rules that were programmed into the system and let the system learn the rules of the game based on the results, creating AlphaGo Zero, which learned exclusively by playing itself over the course of millions of games in just a few days. According to Google, "AlphaGo Zero also discovered new knowledge, developing unconventional strategies and creative new moves that echoed and surpassed the novel techniques it played in the games against Lee Sedol and Ke Jie."[150]

In a board game like Go or chess, it's easy to see why an AI would want to run the same simulations over and over with only minor variations: The system wants to play itself to learn by making different choices and optimizing the results. Similarly, in a quantum circuit optimized to solve a particular numeric problem, you can lay out the criteria with which to use quantum parallelism to find an optimal solution, which means you are essentially doing this in parallel.

Could such methods be used for problems that aren't simply numerical? It turns out that you can use the same approach for any scenario that can be simulated using bits and bytes, even graphical worlds like video games. That was the point of my previous book, *The Simulation Hypothesis*. Three-dimensional worlds are ideal for playing and exploring when the idea of optimization is not as well-defined as it is in a simple board game.

The process is already underway, as is using these simulated worlds to train AI. Self-driving cars, for example, must be trained in the real world as much as possible. But this is not only an expensive proposition; it becomes downright dangerous if you are trying to train self-driving algorithms for edge case scenarios, such as when a person walks out in front of a moving car. Many of these cases may lead to a crash, which is much less costly in a simulated 3D world than in our physical 3D world.

A system developed at MIT, called Virtual Image Synthesis and Transformation for Autonomy (VISTA), projects a small dataset of a particular location and circumstances, which were captured from a real driver, inside a 3D model of the world to the self-driving algorithm. It can present variations of the data and test the algorithm by changing parameters or render it from different perspectives.[151] In 2020, Elon Musk announced a self-training computer at Tesla, called Dojo, that uses massive amounts of video footage to train the self-driving cars.

Self-Driving Cars and Fitness Functions in 3D Worlds

Self-driving cars are an interesting bridge between a simple fitness function for a simulated game, such as that for chess or Go, and a more complex simulated environment that might have multiple goals. For example, the self-driving algorithm wants to get to the designated destination fast, but also wants to follow all the rules and do so in a way that doesn't endanger the occupants or pedestrians or other cars.

If you could think of a simulated environment with multiple agents—let's say self-driving cars—that were all training themselves, you would get closer to this idea of rerunning scenarios many times to get better outcomes in a massively multiplayer simulation. Each agent or player has similar goals or might have different goals for a particular run of the simulation, but the environment is shared.

In a role-playing game, particularly a multiplayer role-playing game, the fitness function, which tells us how good one outcome is compared to another, is much more subjective and will depend on the design of the game and players. If we are in an NPC simulation, the reason to run alternate scenarios is very similar to AlphaGo's version of self-play or a simulated environment for training cars. Even if we are in the RPG version of simulation, where we all exist outside the simulation, there must be some complex fitness function that shows the suitability of scenarios for each player, who may have individual quests, goals, and experiences they want to have in the game.

In addition to an individual fitness function (How good is this simulated timeline for me? or How did I do?), there might be collective fitness functions (How good is this timeline for a group of people?) or even a universal fitness function (How good is a timeline for everyone?).

If, for example, JFK is assassinated, it may not be a good outcome for him in particular in that one timeline, but if another timeline when he is not assassinated ultimately leads to a nuclear war, the individual fitness function for him might be better in this disastrous timeline (he lives longer), but the overall fitness for all the other players is obviously much worse if everyone dies in a nuclear war.

This gives us at least a framework in which to think about multiple iterations of a particular simulation and why the Core Loop might be running in the first place: It is exploring the garden of forking paths, trying to determine the optimal outcome for each individual, who is making all the choices and observations, as well as for the optimal outcome or even likely outcome for everyone, that is, for all the players in the simulated game collectively.

Tom Campbell and the Fundamental Process

It turns out that some physicists have gone further in their interpretation of a multiverse than others to account for this idea. One such physicist is Tom Campbell, author of *My Big TOE*, which came out in 2003 and is about his Theory of Everything. Campbell, who was a physicist who did work for NASA, has long advocated that we are living inside a virtual reality. Moreover, he may have been one of the first to articulate fully how the universe was constantly "branching out virtual realities" as a fuller explanation for the quantum measurement problem.

Campbell calls this the Fundamental Process, and he has two versions of it. One version is for inanimate objects such as particles and the physical universe. The Fundamental Process spins out virtual realities to explore every possible process and then chooses the best one based on what is most profitable. For the universe and particles, Campbell defines what is most profitable as that which requires the least amount of energy, making it the cheapest and easiest to achieve. It also results, as physicists like to say, in greater disorder over time in the universe, or greater entropy.

For conscious entities, however, like humans and other living beings, the definition of most profitable is different. Campbell defines it as that which reduces entropy, moving toward greater order and *less* entropy over time. Campbell writes:

> The Big Picture Fundamental Process of evolution (or fundamental process for short) is as follows. An entity starts from any point (level) of existence or being, spreads out its potentiality into (explores) all the available possibilities that are open to its existence, eventually populating only the states that are immediately profitable while letting others go.[152]

At the time I first read Campbell's book, I was looking for more concrete evidence that we live in a simulated reality. I didn't devote a lot of cycles to consider the full implications of the universe branching off different virtual realities and making choices. In essence, his Fundamental Process is very similar to the Core Loop idea articulated in this book and, to a certain point, similar to the process described by Philip K. Dick in the first chapter of this book.[153]

The Future Sending Messages to the Past

Another alternative interpretation for the quantum measurement problem is that several possible futures exist, and they are sending information back in time to what we call the present. Although this explanation sounds like science fiction, it is not entirely inconsistent with our ideas that multiple realities are playing themselves out and that what we think of as a single timeline is actually multiple timelines we are choosing from. Nor is it, when you think about it, inconsistent with the delayed-choice experiment results; it's simply stated from a different point in time (what we call the past in the delayed-choice experiment).

This explanation, based on the work of John G. Cramer from the University of Washington, is given by Fred Alan Wolf in his book, *Parallel Universes*.[154] Cramer pointed out that what the mainstream interpretation calls probabilities of certain futures can be calculated by combining two waves, one that moves forward in time (the standard probability wave of quantum mechanics), which he calls an offer wave, and one moving backward in time, which he calls the echo wave. As has been pointed out, the equations of quantum mechanics work when time is a negative number, and this provides a mathematical solution where multiple futures exist and are sending information back in time to the present. When you multiply the offer wave with the echo wave, this is how you end up with the probabilities that are discussed in the standard Copenhagen interpretation.

Wolf provides an interpretation and addition to Cramer's ideas that tie it to the multiverse. In Cramer's original approach, just one possible future is sending back the echo wave; this ends up being the one that is most probable—that is, the measurement that we make. Wolf goes further and asks where the other probabilities would then come from. Wolf says: "All futures return the message, not just the best chance future."[155] For these possible futures to send back information, they would have to exist. For them to exist in the future, there would have to be some path to get there. This again defies common sense and even most classic understanding of time, but it is consistent with the idea of the delayed-choice experiment and a simulated multiverse.

When we introduce the simulated multiverse and the Core Loop and quantum computing, this seemingly nonsensical idea makes much more sense.

But this begs another question: What does it mean for a future to exist?

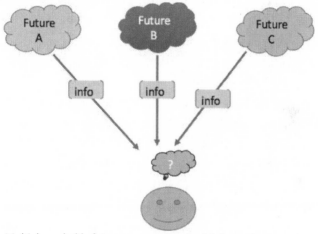

Figure 53: Multiple probable futures are sending back information we use to make decisions.

In a simulated multiverse, each possible future is played out, assigned a probability of some sort or a fitness that determines which is the best possible future for us to run. It's not much different than a simple AI or a simple quantum circuit; the superposition is used to run each scenario in parallel. Are they really futures, as shown in Figure 53, or just different runs of the simulation? It turns out there is no difference, at least in the simulated multiverse!

Principles of a Simulated Multiverse

Before we move the discussion outside of the realm of science and computing and consider what it all means to us as individuals and from a spiritual perspective, let's recap the principles of the simulated multiverse as we have been exploring them throughout this book.

1. The physical universe is based on information.
2. The physical universe is rendered based on information.
3. The universe is a computing system.
4. The future is a computationally irreducible process.
5. The universe consists of many possible futures, each of which runs as a separate process.
6. What we think of as time doesn't exist in the traditional sense.
7. What we think of as the fixed arrow of time is actually a series of line segments, going from one gamestate to another.
8. What we think of as the past is the saved gamestate/memory of the past, not the actual past.
9. There can be multiple possible pasts. There can be multiple possible futures. These are expressed in paths through the Multiverse Graph.
10. The universe is constantly projecting to the future, creating multiple timelines, branching and merging and pruning. This is what the Core Loop is all about.

RUN LOLA RUN

In *Run Lola Run*, a German film released in 1998, we see a kind of multiverse, or multiple runs of the same scenario, playing out in front of our eyes. With very little exposition, we meet Lola who gets a desperate call from her boyfriend, Manni. Manni has been involved in some underworld activities and he is going to be killed by some shady mafia types unless she can deliver 100,000 marks (pre-Euro currency in Germany) within 20 minutes. If that isn't stressful enough, the choices that she makes seem to involve life or death not just for Manni but for Lola as well. We see her making strange choices—like going to her father, who is a banker, and holding him up at gunpoint for the money. Each time, we see either Lola or Manni being killed at the end of the segment.

When that undesirable outcome happens, the film rewinds (complete with Lola going backward) to the same starting point, and the process starts again. Lola goes forward and makes different choices, which lead to a slightly different and, hopefully, more optimal timeline. Technically, these choices could be considered parallel worlds from the quantum point of view, but they are seen as temporary timelines, which are then erased when they lead to a less than desirable result (as most of them do).

In this film, we literally see our Core Loop being played out, starting with a moment in time when Lola gets the phone call. We see the state saved somewhere, a set of choices made, and a new timeline emerge; we then literally see the rewind back to the starting point, and Lola is allowed to make different choices to see what the results would be. There is even a clear way to evaluate each timeline: Obviously if Lola and Manni both live, this is considered the best timeline.

Are we all Lolas?

Chapter 14

Stepping Back—What Does It All Mean?

"We realize that the universe bifurcates in every such event in the transcendent domain, becoming many branches, until in one of the branches there is a sentient being that can look with awareness and complete a quantum measurement."

-Amit Goswami, *the Self-Aware Universe*

The Purpose of Simulations and the Multiverse

Thus far, this book has mostly been focused on scientific concepts (physics, computer science, video games, automata, quantum computing), with science fiction as inspiration. These have been the tools we've used to explore the idea of a simulated multiverse—a virtual garden of forking paths. In this chapter, I want to zoom out and explore some of the bigger questions that inevitably come up when talking about living in any type of simulated universe and that only multiply when living in a simulated multiverse.

After I had published *The Simulation Hypothesis*, whenever I did an interview or presentation, two of the most common questions I would be asked were:

1. What is the purpose of the simulation?
2. What does this mean to me? How does it affect my life?

You'll notice that these questions move beyond science into

religion and philosophy, and thus the answers to these questions have to be both speculative and subjective. My own answers tend to bring forward some of the material about why religions and simulation theory are not so distinct as they might first appear, though ancient religions used completely different metaphors than we use today. Moreover, I felt that the metaphor of a video game-like simulation (RPG version) was one that could bring the worlds of science and religion closer together, because they seem to have drifted further and further apart. I believe the first book accomplished that to a certain extent, and I now have no problems discussing the simulation hypothesis with colleagues in science, industry, traditional religions, or those in that ever increasing classification of "spiritual but not religious."

In this chapter, I'd like to try to step back and see whether we can get a similar perspective for a simulated multiverse.

Amit Goswami and *The Self-Aware Universe*

One physicist who takes the problems of quantum mechanics out of the realm of the physical to talk about probabilities and a transcendent realm is Amit Goswami, a professor of physics at the University of Oregon. I met him briefly several years ago in Mountain View, and at the gathering I remember someone asking him about the wave-particle duality and the observer effect.

His answer, which I now recall only dimly, was something along the line of: It isn't really a problem per se; the probability wave is what happens with multiple particles—if you were to run the same process again and again, you would see the pattern of probabilities.

I only vaguely remembered what he had said and decided to look in more detail at his views while writing this book. It turns out that his views were closer to the Copenhagen interpretation than to the MWI, but I found an interesting aspect of how he brings both interpretations back into the realm of consciousness while being relevant to our topic of a simulated multiverse.

Goswami argues, as I have done in both my previous book and this one, that the idea that the universe doubles the amount of physical matter at each instant seems to offend the idea of scientific parsimony. A better possibility is that each split is really a potential split, and this process continues and solidifies as individual choices are observed. Goswami says:

> Instead of saying that each observation splits off a branch of the material universe, we can say that each observation makes a causal pathway in the fabric of possibilities in the transcendent domain of reality.[156]

What is this transcendent domain of reality? Goswami's answer is that this is the transcendent entity called consciousness that is spoken of in many religious traditions. But like the magic of infinity, at this point, he has gone into a realm that science cannot effectively make any statements about or give a cohesive framework for. Goswami's explanation sounds a lot like Campbell's, who refers to pure consciousness outside of what he calls PMR, or physical matter reality.

But this transcendent realm also sounds a lot like Philip K. Dick's ideas of orthogonal time, a place outside of what we call linear time. In Philip K. Dick's version, there is a Programmer and a Counter-Programmer who are making changes to the "chessboard of the universe," exploring the different results of different timelines, which are arranged laterally like "suits in a closet."

In both cases, the simulation hypothesis gives us a better framework for understanding the place of our universe in this transcendent realm. What Goswami calls an observation is actually a particular rendering of one of the possibilities in the very large set of possibilities. By considering that these multiverses are probable realities, and only running the ones that we need to get the answer, we are then able, as both Goswami and the Copenhagen interpretation indicate, collapsing the probability wave to a single possibility. However, this might also be the most actuated possibility, as Philip K. Dick would call it, from our point of view at this point in time, but it is not the only possibility in the present, past, or future.

Revisiting Simulation and Religion

The first big question I like to ask my audience to consider when talking about any kind of simulation is whether we are in an NPC versus RPG version of the simulation.

In the NPC version, we are all AI built with a specific purpose, and we don't necessarily have free will, though there may still be an element of randomness if the simulation is running using quantum principles.

In the RPG version, we are controlling or role playing a character (our avatar) that has certain attributes—not unlike a *Dungeons and Dragons* (or *World of Warcraft* or *Fortnite*) character—with different strengths, weaknesses, proclivities, desires, and so on. But we are not our avatar, we are simply inhabiting the character and have (at least from the point of view of the simulation), free will.

The RPG version of the simulation hypothesis has been called out by many as being akin to religion in a way. In *The Simulation Hypothesis*, I made it a point to have a section outlining similarities between various religions and what it would mean to be in a simulation. The ancient texts used metaphors that could be widely understood at the time (books, stage plays, dreams, rivers, supernatural beings). The implications of these metaphors were quite clear: that we are not living in the real world—that is, this is not all there is. This is most particularly true of the Eastern religions (Hinduism and Buddhism), but I found it was also true with the Western religions (Christianity, Islam, Judaism) and their ideas of an immortal soul, angels, and an eternal afterlife presided over by a supreme deity.

I would argue that the simulation hypothesis is the latest update to the metaphors that have been used by the world's religions and their founders, and that these metaphors should be updated to reflect terms relevant to the next generation, including computers, television shows, iPhones, and, of course, what I think is the most important and relevant of all the metaphors to date: video games.

In just one example of such an attempt to update the metaphor, the Mormon Transhumanist Association has put forth the *New God Argument,* which argues that we are in a simulation of some sort and that this is not in conflict with their faith (the Church of Latter Day Saints). This is an explicit attempt for existing religions to stay on top of the theological implications of the development of technology and the simulation hypothesis.

One of the points that Bostrom makes on the simulation argument website is that he has come across many hardcore atheists, who, after considering the simulation hypothesis, changed their view to agnostic. In fact, the simulation hypothesis has been called religion for atheists. David Pearce, a British transhumanist who influenced Bostrom, puts it another way that I agree with: "The Simulation Argument is perhaps the first interesting argument for the existence of a Creator in 2000 years."[157]

Metaphors, Souls, and Illusion

Bridging the gap between the Western and the Eastern religions is the idea of an immortal soul, which comes into the body but is not the body. This is perhaps one of the fundamentals of most religions and is at the core of the religion-versus-science debate. The materialistic point of view defines the self as the body, and the religious point of view defines the self as the soul, or consciousness that exists independently of the body and goes on after the body dies.

The RPG version of the simulation hypothesis also says that we are not our bodies, but that our physical bodies are our characters, whereas the real self is actually the player, not the character itself. The Greeks, who had their own religion, believed that we crossed Lethe, the River of Forgetfulness, when we incarnated, and other cultures have had their own versions of forgetfulness.

When matching the simulation hypothesis to the Hindu and Buddhist concepts of illusion and reincarnation, one barely needs to draw an analogy or speak in metaphor. The ancient texts were clearly using older metaphors. One of their core tenets is that the world around us is *maya*, a Sanskrit term that roughly translates to "illusion" or, with more nuance, "a carefully crafted illusion." We are led to think it is real, it feels real, but it is, in fact, illusory, made only of light.

Another metaphor often invoked in Buddhism is the idea that world is like a dream. The Hindu Vedas, the oldest written scriptures of a major religion, speak of the *lila*, or the grand play of the gods, as if we were participating in a stage play as actors, a metaphor picked up by Shakespeare in the Middle Ages when he wrote, "All the world's a stage ... And all the men and women merely players."

Coming forward to the past century, Paramahansa Yogananda, author of *Autobiography of a Yogi*, widely acknowledged as one of the top spiritual books of the twentieth century, writing in the middle of that century, updates these to a more modern metaphor: the world is like a moving picture seen on a screen at a movie theater. Everything is a carefully crafted illusion and we get so caught up in it that we forget there is a beam of light from the projector to the screen that is giving us the illusion of motion. This is a good metaphor because there is no such thing as motion in a motion picture; it is just a carefully crafted illusion.

If Yogananda or the rishis (shadowy figures in history who wrote the Vedas) or even Shakespeare were alive today, I believe that they would use a more up-to-date metaphor, one that combines a stage play or film with free will, scripts that can change based on choices made by individual players. Basically, they would say that we are living in an interactive video game.

In a video game, the world consists of pixels in a very carefully crafted illusion. You could imagine a video game Buddha telling other characters in the video game, echoing the words of the historical Buddha:

> *Know that all phenomena*
> *Are like reflections*
> *Appearing in a very clear mirror*
> *Devoid of inherent existence.*

Gods, Angels, Demons, and the Programmer

To those of us in the simulation, anyone outside of the simulation would seem like supernatural beings—perhaps God or gods or devils—a trend echoed by Philip K. Dick in his speech at Metz, where he used the terminology of a Programmer and Counter-Programmer. Watching the full speech and his other writings, you can see these terms are obviously religiously loaded. Even Bostrom mentions a naturalistic theogony in his original paper, pointing out that the creators of the simulation would seem like supernatural beings or gods to us because they would effectively be omniscient.

In the Western religious traditions, what exists outside of physical reality? There is more than just the Creator who says "Let there be light!" in the Old Testament. There are also angels that are depicted as messengers of God. In some cases, these angels are delivering messages to humans; in others, they are the instrument of God's punishment, and in some cases they are recording angels, recording our every action and deed so that we can be judged on those.

Sometimes they are more like automata or, using the Greek term we use in computer science, *daemons,* just carrying out their assigned duties sans personalities or choice. Most religious traditions tell us there is a world that we cannot see that is populated by beings of various spiritual (or demonic) levels, and that they are watching us. Sure sounds like we are in a fishbowl of some kind. Another way to say it is that we are in a simulation where these superusers are logging in and/or launching semi-autonomous processes with specific purposes that have more access privileges than we do, but are each constrained by their purpose.

Reincarnation, Karma, and the Scroll of Deeds

In the Eastern traditions, our soul downloads into a body and we play that character for our entire lives until bodily death. There is something that, in addition to the soul, survives bodily death: our karma. We then download later to another body, we identify with that body, or character, and we carry the information about what happened (the karma generated and resolved).

Where is this information stored? I would suggest it is in some kind of invisible cloud data storage. In the famous Buddhist Wheel of Samsara, we are caught in maya, going round and round, playing the game again and again, learning lessons along the way, resolving our karma. You hardly need to draw an analogy of a video game being rendered based off the information stored on a server. The players enter the game playing different roles, where each has a set of quests and achievement for each character or life. I outlined this in great detail in my previous book.

In fact, it's not just in the Eastern religions that something is keeping track of all of our actions and recording them. In the Western traditions, these are recording angels; in Islam, there is a name for them, *Kiraman Katibin*, who write down everything on a Scroll of Deeds. Obviously, this would be a big scroll if it contained all of our deeds in one lifetime. Imagine having to do that for seven billion people! The Scroll of Deeds is analogous to the Book of Life in Christianity and Judaism, which is often presented as a list of names of who gets into heaven and who doesn't. But in some cases, it contains a list of deeds, not just names.

In the Eastern traditions, there is the idea of karma, of unfinished business, which is not so much about getting into heaven or hell but about what your next life (or lives) will be about. In a sense your future lives are constrained by choices you make in this life. Enlightenment is reached when you are able to deal with all of your karma and not create new karma. Both of these lend themselves well to the video game metaphor and a computer system or database of something that is keeping track of all these things that happen.

Near-Death Experiences and the Bardo

What evidence is there that an area outside the simulation, that is, outside of our physical reality, exists? Personally, I tend to pay attention when people claim to have been outside, especially if there is a consistency across reports. One area where we have a lot of consistency is in the reports of thousands of people's near-death experiences (NDEs), many of whom experience a life review, a blow-by-blow replaying of everything that happened in their lives.

I was first exposed to this through my friend, Dannion Brinkley, who wrote the bestseller *Saved by the Light*, about his own NDE after being struck by lightning in 1975. One of the areas that many NDEers tell us about is a three-dimensional holographic-like projection of the events of their life, which is a much more colorful and detailed way of talking about the old phrase, "My whole life flashed before my eyes." Dannion refers to it as a panoramic, 360-degree replay of your life from any (or every) angle.

This life review isn't done purely for mechanical purposes, NDEers tell us; rather, it's to let us examine the choices that we made and how they affected other people. In fact, in addition to its realism and panoramic, three-dimensional projection, perhaps the most defining characteristic is that you get to see what happened from the point of view of other people in your life, particularly those that you made feel good or perhaps even hurt or wounded (intentionally or unintentionally).

Although our video games do not have the ability to replay emotions, you can imagine in the future, with BCIs (brain-computer interfaces) and 3D projection, that we would be able to record and replay not just visual cues but also feelings and sensations that were players experienced. In fact, one of the startups that I was involved with in Silicon Valley replayed any scene from a 3D MMORPG (for example, *Counterstrike: Global Offensive* or *League of Legends*) in virtual reality, so you could put on a headset and experience a panoramic, three-dimensional holographic replay of any event from any point in the virtual space, or from any character's point of view.

If we take reports of people that have been beyond the veil at face value, it seems like whoever is running the simulation already has this technology in place, with every scene being recorded from every person's point of view, and is then playing it back for a coaching session, where we look at our choices. Presumably, just as it's more efficient for us to run millions of simulations of a game like Go inside the computer to train AIs, whoever is running our simulation finds it easier to train us by going through these types of experiences and then reviewing the results with us. In traditional professional sports, it is common for a coach to watch a film of the previous game and get individual athletes to review what they have done to try to play better next time. In more modern competitive sports, such as eSports, this is literally watching a recording of a 3D game that had multiple players, in order to do better next time—which, it turns out, is the purpose of the life review as well.

Previews of Future Events

In most reports of the life review, the events shown are past events, which would fit into a more traditional idea of time (though with a twist, since these theatrical events occur outside of time and space, so there must be another time and space that is, to use Dick's term, orthogonal).

Do any NDEers report actually seeing future events? This, I suppose, would be a life preview and not a life review. Actually, there are a number of subsets of NDEers who report seeing not just future events but future possibilities. Many NDEers are presented a choice of whether to go back to their lives on Earth or to stay in this blissful world they find themselves in. Most NDEers report not wanting to go back, because the world that is beyond the physical (or, in our terminology, outside the simulation) doesn't have the pain and fear and many of the negative aspects that exist in our world.

Why, then, do they come back? Some aren't given a choice; they are simply told it isn't their time. Others make a conscious choice to return, but usually only when they are shown future films of what would happen if they didn't go back. One woman reported seeing her children growing up without their mother and how this led to various paths that weren't ideal in their lives. It was as if she could see their lives being played out if she made one choice—kind of like our alternative timelines—and then came back to the present to make a choice about which of the possible timelines might actually proceed.

Another very colorful example of this was Natalie Sudman, an Iraq War veteran who had a NDE as a result of an explosion of an IED. Recounted in her book, *Applications of Impossible Things*, and in various interviews online, Natalie gives a very thorough account of her afterlife experience. When she discusses the life review, she makes the point that it wasn't just a series of events flashing before her eyes, but rather that "a simultaneous awareness exists of the interconnectedness of innumerable strings of being, expansion, possibilities and probabilities, back in time and forward, sideways and between what we perceive from the physical."

Sudman continues by telling us that this process "...includes alternative paths of action, reaction, and interaction as they may have played out had I chosen this or that option instead of what action I did choose to follow at any point of physical life." [158]

Sudman goes on to say that because this includes exploration of possibilities, it might more appropriately be called a *life exploration* rather than a life review.

The place in the afterlife where these life reviews are conducted seems to be like a super-duper entertainment room where you can not only fast-forward and rewind the movie, but go sideways, like an interactive simulation, to see what *might* have happened. It's as if there are many storylines, like an interactive choose-your-own adventure at each moment in time. This relates directly to our idea of the Core Loop, where we are able to explore possible alternative timelines and see *what might* happen, given a certain change in variables.

After that, and perhaps even more startling, she was shown what was going to happen next in her life after she went back, but this also had different parameters that might lead to different timelines. In her case, the different explorations were the various injuries she might suffer in the IED explosion and what experiences she might have when she went back into her life with those injuries. This might seem like a horrific episode to some (visualizing what different injuries might occur), but Sudman tells us that it was the exact opposite; it was like having fun planning what might happen if she couldn't walk or how other people might react if she sustained or did not sustain certain injuries. For example, if she had become totally blind, how would people react? If she was completely healed and suffered only minor injuries, how would that make other people feel? In the end, she did sustain a number of injuries, but they mostly healed over time. It sounded a bit like actors and a director discussing different scenes in a script and deciding which ones would be best for the story of the film.

Life Preview and Life Planning

Research by Dr. Brian Weiss (known for his book, *Many Lives, Many Masters*, and several sequels) and Dr. Michael Newton (best known for his book, *Journey of Souls*) seems to cast more light on this phenomenon of being in the space between lives. The Tibetans call this place the *bardo* or the *in-between*.

In some cases, the clinical evidence comes from patients through hypnosis and, in some cases, through meditation or spontaneous recall of a time before they were born, or their time in the bardo. What makes these accounts interesting is the commonalities that are found, just like the commonalities of NDE reports.

Dr. Weiss writes in *Messages from the Masters* of what we would call the life review in the bardo:

> There is considerable evidence that we actually see the major events in the life to come, the points of destiny, in the planning stage prior to our births. ... Some cases of déjà vu, that feeling of familiarity, as if we have been in that moment or that place before, can be explained as the dimly remembered life preview to its fruition.[159]

Here we see another esoteric explanation of the phenomenon of déjà vu, this time backed up by clinical evidence. Though not rising to the same level as physical evidence, it suggests that Dr. Weiss and others are tapping into the fabric of reality in a way that science doesn't understand (at least not in the limited view of materialism). Like Philip K. Dick before him, but this time relying on the descriptions given by his patients, Dr. Weiss and other pioneering therapists may have stumbled upon the spiritual explanation of our simulated multiverse, where scenes are played before we actually experience them, perhaps played out with different possibilities based on major choices, resulting in branches of timelines.

An even more vivid description of this part of the bardo is given in Dr. Newton's groundbreaking book, *Journey of Souls*. Dr. Newton conducted hundreds of regressions that took the patient not to a previous life but to this in-between place. In it, he describes a number of virtual places that patients live in while in between lives. These descriptions sometimes have analogs to the descriptions by NDEers, such as the presence of beings of light as guides who are there to help them analyze their mistakes during previous incarnations (or, as I would say, with their previous characters in the simulation).

Dr. Newton goes further to describe this idea of planning out timelines during a period of life preview. One of the more unusual parts of Dr. Newton's findings was that of life selection and planning, which he says takes place in a location that has a special control room to preview possibilities. Dr. Newton writes: "I am told it resembles a movie theater which allows souls to see themselves in the future, playing different roles in various settings."[160]

Dr. Newton describes conversations with patients who entered the room and sat in front of many screens, again like an entertainment center or, as one patient described it, "It's as if someone flipped a switch on the projector in a panorama movie theater. The screens come alive with images and here is color ... action ... full of light and sound."

This patient goes on to describe a scanning device in front of the screens that is full of lights and buttons as if he were in the cockpit of an airplane. When he describes operating the scanner (or projector, depending on which metaphor you like) through this control panel, he says he starts watching what looks like a three-dimensional movie of his *upcoming* life in New York, but it's not a movie; it's as if he was *actually* watching life going on in New York from outside, even though it is an event in the future.

This particular patient (and many others) describes "lines converging along various points in a series of scenes," as if the operator of the machine is traveling through time, watching his upcoming life unfold. This patient and others describe major decision points in their lives, which is where the timelines converge and diverge, depending on what decisions are made by their upcoming future self, and each decision point branches a possible timeline. Moreover, this patient, who is perhaps more descriptive than many others but is consistent with other patients, says that it is both like "watching a movie" and "jumping in and seeing things from anyone's perspective in the scene." The controller can be used to move forward or backward in time and along the many lines of alternatives that are played out.

If this is true, then whatever is involved in controlling this simulation could do the same thing for billions of other souls who are going to jump in and join the massively multiplayer simulation. Each has control of viewing what would happen with different choices, at least with major decision points that could create alternate timelines. What's interesting is that these can be played out when they are still in the bardo before birth, or after death, reflecting possible pasts (in the life review) as well as possible futures (in the life preview).

What are we to make of these very science fiction metaphors, of movie projectors with controls moving time forward or backward, in the middle of what seems to be a spiritual place, the afterlife or the bardo?

Alternative Lives and Parallel Realities in a Spiritual Simulated Multiverse

Although most traditional religions don't talk about multiple timelines, clinical evidence from NDEers and from those who remember planning out this life suggests that there is a mechanism, somewhere outside of the physical world, for previewing timelines in our lives, complete with branches that can be played out from a type of monitor to watch the consequences of different actions.

These timelines are created by our characters making different choices. We are then not only able to see the scenario play out just in general terms but actually watch what might happen through some kind of technological control panel. Some even say they watch what is happening from outside the physical world.

What does that sound like? An RPG version of a video game. Our world is what we have been calling a simulated multiverse, which can run forward and backward for many players. Where does all this occur—that is, what is outside the game? We don't really have a good term for this, so I have borrowed the Tibetan Buddhist term, bardo (though afterlife or beforelife would be just as appropriate).

This bardo seems similar to the Philip K. Dick idea of orthogonal time and Amit Goswami's ideas about a transcendent domain.

In fact, given that the timelines can branch, it starts to resemble more of an interactive video game with a sense of quantum computing and quantum parallelism, where we spin out different worlds to explore alternatives and find the best ones by playing them out. After a life, we can review, watching not just what had happened, but see alternative timelines that might have happened had we made other choices. This means that all of the timelines must have, to some extent, happened in some version for us to be able to watch them. Another way to say is that they can be rendered based on specific inputs of actions at any point.

These timelines are in fact part of the complex pattern of the garden of forking paths that is our lives, meshed with the complex pattern of other lives, viewed from a vantage point where all the gardens are visible.

Although most scientists will dismiss this spiritual explanation of multiple timelines based on our choices, partly because there is no way to validate or invalidate it, I think this provides yet another dimension to our exploration of the idea that reality consists of more than one timeline. The branches of timelines described by those in the bardo watching physical reality seem to be primarily concerned with major life choices—will you go to college on the east coast or the west coast, who will you marry, which career you will choose —and not so much little choices.

What if this is occurring not just at the end of a life or at the beginning of a life, but while we are constantly waking up outside the game in the bardo to try out different scenarios and then pick the optimal one for us to learn our lessons? What if we are still at the controls each night, planning out what might happen next, and then making choices to guide the game along one timeline or another? Then we cross back over the River of Forgetfulness when we wake up?

We now have a spiritual version of the simulated multiverse— the Core Loop operating on a Multiverse Graph— tailored for each player. Basically, we have an RPG version of a massively multiplayer online role-playing game with versions of our Multiverse Graph and Core Loop.

Normally, the mechanisms of quantum physics and the idea of a quantum computer seem like cold, unreasoning processes that go on by themselves like a great big machine.

However, with an added spiritual perspective, we can assign some meaning to the problems that are being solved. Just as Philip K. Dick supposed that there was a Programmer changing variables and watching what happened, the RPG version of the simulation hypothesis gives us an even stranger but ultimately more fulfilling vision: that we are the Programmers, to some extent, all making choices, and the impersonal machine is simply the game engine that is rendering what is happening after our choices, keeping track of all of our characters, and creating consistencies across these different timelines?

This not only explains the Mandela effect, but also gives us a different, perhaps more meaningful, explanation of the simulation hypothesis, complete with parallel universes being driven and rendered by some form of quantum computing engine, bringing together all of the major threads we have been exploring in this book.

BLACK MIRROR AND A DIGITAL AFTERLIFE

Although many of the episodes of Black Mirror tend to overlap with the topics we are discussing in this book, one in particular stands out when it comes to simulations and the afterlife: San Junipero.

In this episode, we start off in the landscape of a beautiful ocean-side beach town, San Junipero, which seems to be stuck in a constant state of partying during the 1980s.

One of the characters, Yorkie (Mackenzie Davis), meets the fun and playful Kelly (Gugu Mbatha-Raw). After an awkward night of dancing and sexual proposition, they meet again next week and start a virtual relationship, despite Yorkie saying that she was engaged to be married. Kelly confesses she had once been married, too.

The following week, Yorkie searches for Kelly in different era nightclubs from different decades and finally catches up to her. Then we learn that San Junipero is a simulated reality for the elderly who are hospitalized so they can revisit their youthful days and forget about the infirmities of their body. This is done with two little sensors that are placed on the forehead and project information into and from the brain, acting as the BCI to the simulation.

In the virtual world, your avatar can be as young as you like, and everyone seems to be in some prime sense of shape in their 20s and 30s. The world feels entirely real while you are plugged in. Not unlike in The Matrix, you are temporarily unaware of your body in rl (real life). There is a catch: If you aren't a permanent resident of the virtual world, you can only log in for so many hours at a time.

This is a common restriction in recent science fiction that uses fully immersive, simulation-level BCIs.

Becoming a permanent resident means permanently uploading your consciousness to the virtual world, which means you are no longer alive as a biological being—your physical body is dead. At that point, even though to you it seems like a digital afterlife, you are actually now part of the computer system, so you would technically be an NPC to those who are visiting the system, though one that was based on the state of your brain at the moment of your physical death.

It turns out that Yorkie is in an elder-care facility in San Francisco and has been paralyzed for life. She is getting married just before she dies to someone at the facility for legal reasons so that he can authorize her uploading as a permanent resident (i.e., her euthanasia).

Kelly visits Yorkie in real life. Yorkie wants Kelly to join her as a permanent resident after she dies, but Kelly is hesitant. Kelly believes she will see her dead husband and her daughter in an actual spiritual afterlife, since the daughter passed away before uploading to a simulation was possible, and her husband had chosen a natural death as well.

San Junipero depicts an idyllic world of partying and never-ending fun, with beautiful landscape by the ocean in Mexico, a different take on virtual worlds than in *The Matrix*. It takes our current 3D worlds where we can have avatars and can design our own residences to a logical conclusion, but it raises some interesting questions about what happens after we die and the nature of consciousness itself, as well as the nature of a digital afterlife versus a spiritual afterlife.

What Does All This Mean for Me? Does It Matter, Really?

I'd like to end this book on one of the questions that I started this chapter with: What does this mean for me, the individual, in living my life within a simulated multiverse?

Before we try to answer this question for a *multiverse*, first let's start with what it might mean to live in a single simulated universe.

My personal perspective is that it obviously depends on whether you believe we are in an RPG version or the NPC version.

Believing that we are all NPCs can have a bit of a nihilistic effect (nothing that I do matters!). That said, if we are all NPCs, there must be some bigger reason for simulating this universe, just as there are reasons for running any simulation. Even if the outcomes that the simulators are trying to measure or determine are global, the actions of individual agents within those global goals do matter: they add up to the global results. An NPC simulation might be trying to answer big questions like: Will the civilization blow itself up? Will they ruin the planet? or Will they leave the planet?

The first version of this book was published in 2021, which is looking like the year that will have the most simulation-themed movies since 1999. One of the movies released this year is called *Free Guy*, and it is about an NPC played by Ryan Reynolds, whose name is just Guy, in a video game city that is scheduled for shutdown. Guy decides that he's going to do something about it and starts to make a difference to him and other permanent residents of his virtual city. Although this plot might seem farfetched to anyone who programs video games today, the message is that the actions of every individual agent within a multi-agent system can matter and shouldn't be lost.

On the other hand, if you tend toward the RPG version, as I personally do, then it's much easier not only to take this revelation in stride but for it to have a positive effect on our lives. We play video games to have experiences that we might not be able to have outside of the game (such as flying on a spaceship or riding a dragon). Video games are no fun without challenges. As Nolan Bushnell, the founder of Atari, used to tell his game designers: Make games that are *simple to play but hard to master*. These are the games that have staying power.

Similarly, if you take that point of view in the Great Simulation, you can see that life as we know it is easy to play, but hard to master. In fact, to keep things interesting, we need challenges, or the game will become boring. In the original *Matrix* trilogy, for example, it was revealed to Neo that the original version of the matrix was a blissful experience, but that the humans didn't take to it or accept it as real. The word *utopia*, in both Greek and Latin, literally means "nowhere."

The truth is, we all have challenges at some point in our lives, whether we are facing racism, health problems, relationship issues, family issues, physical violence, money problems, emotional issues, not to mention loss of loved ones and so on. Each of us, like a video game character, has certain strengths and weaknesses, and for whatever reason, our challenges are also often tailored uniquely to our personalities.

Personally, I like to think about the challenges in our life as if they are quests or achievements within a video game. Not only do we have to keep trying to overcome these challenges to level up; we also at some point have to accept these quests before they appear on the timeline of our character. In a sense, if we are in an RPG simulation, we may have chosen at least some of our challenges.

Sometimes the difficulty level of these challenges is small. At other times, it seems so high that it feels insurmountable. Sometimes, we need other characters to work with us to overcome these challenges and can't do it alone, like a task for a guild in a video game. But that doesn't mean we should give up. Would you give up trying to level up a character in a video game because you didn't overcome a challenge the first time (or second or third time)?

With that in mind, let's move to the question you may have been pondering while reading this book: What would it mean for us to live in a simulated multiverse?

We all make choices in our lives, sometimes minor and sometimes major. Whereas the quantum multiverse tells us that every single decision, even the minor ones (shall I have eggs or cereal for breakfast?), might spawn another universe, that is almost too much for our minds to take in. Perhaps it's more appropriate, and better within the point of view of the simulated multiverse, to think only of the big decisions that we make in our life that take us on very different timelines.

If the universe is some kind of quantum computer, and each of us is tapped into this computer as a process running our own Core Loop, then we are constantly trying out these different timelines to see where they might take us and which is the best.

The Road Not Taken

This brings us to two seemingly shocking conclusions, implied by the Multiverse Graph and the Core Loop we have been exploring in this book:

1. That every major decision we might have made, every road not taken, was in fact taken by some version of ourselves (by a time instance).
2. That we (or someone or something) chose this particular timeline over the others for us to experience in this moment.

The best way to illustrate this is to end the book as we started it, with science fiction. In his novel, *Dark Matter*, author Blake Crouch writes of a man, Jason Dessen, who was on track to be a famous quantum physics researcher (in the field that we are concerned about in this book, quantum mechanics). He was working on putting large-scale objects into quantum superposition. Instead of devoting his life to his research, he made a decision to marry his girlfriend (Daniella, who was pregnant with their son, Charlie) and make a life with his family. As a result, he abandoned the research and never became the famous researcher who published and won prizes. Daniella, who was an up and coming artist, also gave up her artwork to focus on raising their child and never became a famous artist.

Though they are both ostensibly happy, in the backs of their minds, both Jason and Daniella had that nagging question about the path not taken. What if they had made different choices? Would they be happier? Or less happy?

In a scenario worthy of a Hollywood blockbuster, Crouch shows us that these other paths aren't always better than the current path. Another Jason Dessen (Jason-2), from a timeline where he didn't marry Daniella and never had Charlie and became a world-famous, award-winning scientist, shows up in this timeline (using his research on quantum superposition to accomplish this feat of multiverse jumping).

Jason-2 then attempts to steal Jason-1's life, trying to replace him so that he could step in and experience the fruits of the road not taken: a life with a family and happiness that was sorely lacking in the other timeline. While the novel goes on taking Jason through many twists and turns, including the odd situations of meeting many versions of himself, there is a point to be made about Jason's current life and for each of us in our current life.

Having taken other roads might have led to other destinations in the garden of forking paths, but we are in this particular one for some reason. There is some aspect of this life that is optimal for us, because a lot of branches of the tree may have been tried, but our player, or the quantum computing algorithm that is running the simulation, has decided what might make us less happy. His experience timeline-hopping gives Jason Dessen a new appreciation for the choices that he did make in this life in this timeline!

And perhaps that is the perspective that I would like to leave you with when trying to answer these big, almost unanswerable questions. If you are in this timeline, don't worry about the other possible presents that may be existing, may have already existed, or may exist in the future. One way or another, you will most likely end up in the best possible one.

And with that encouraging thought, I will end this book echoing Ts'ui Pen: I leave to some, but perhaps not to all possible futures, my garden of forking paths, our shared simulated multiverse.

Suggestions for Further Reading

Bostrom, Nick. "Are You Living in a Computer Simulation?" *Philosophical Quarterly*, Vol. 53, No. 211, 2003, pp. 243–255.

Brown, Julian. *The Quest for the Quantum Computer*. New York: Touchstone/Simon & Shuster, 2001.

Campbell, Tom. *My Big TOE*. Lightning Strike Books, 2003.

Colts, Eileen. "My Mandela Effect Awakening" in *Mandela Effect: Friend or Foe?* Estero, Florida: 11:11 Publishing House, 2019. P14–18.

Crouch, Blake. *Dark Matter: A Novel*. Ballantine Books, 2016.

Deutsch, David. *The Fabric of Reality*. London: Penguin Books, 1997.

Gleick, James. *Chaos*. New York: Penguin, 1988.

Goswami, Amit. *The Self-Aware Universe*. New York: Tarcher/Putnam/Penguin, 1993.

Greene, Brian. *The Hidden Reality*. New York: Vintage Books/Random House, 2011.

Kaku, Michio. *Hyperspace*. New York: Anchor/Random House, 1995.

Kaku, Michio. *Parallel Worlds*. New York: Penguin Books, 2006.

Levy, Steven. *Artificial Life*. New York: Vintage Books/Random House, 1992.

Lloyd, Seth. *Programming the Universe*. New York; Vintage Books/Random House, 2006.

Hans Moravec, "Simulation, Consciousness, Existence", 1998.

Newton, Michael. *Journey of Souls*. St. Paul, Minnesota:

Llewellyn Publications, 1994.

Sudman, Natalie. *Application of Impossible Things*. Huntsville, Arkansans: Ozark Mountain Publishing, 2012.

Sutin, Lawrence, ed., and Dick, Philip K. *The Shifting Realities of Philip K. Dick*. New York: Vintage Books/Random House, 1995.

Tegmark, Max. *Our Mathematical Universe*. New York: Vintage Books/Random House, 2014.

Virk, Rizwan. *The Simulation Hypothesis*. Mountain View, California: Bayview Books, 2019.

Wolf, Fred Alan. *Parallel Universes*. New York: Simon and Schuster, 1988.

Wolfram, Stephen. *A New Kind of Science*. Wolfram Media, 2002.

End Notes

Chapter 1: Down the Rabbit Hole—From Google into the Mind of Philip K. Dick

[1] From the episode, "Blink," 2007 (Episode 10, Series 3, of the revised *Doctor Who*).

[2] The Metz speech has been recorded and is on YouTube in many locations. A full version of the text, was written out as, "If You Find This World Bad, You Should See Some of the Others," and reprinted in many places, including in the *PKDS Newsletter*, no. 27, August, 1991, and published in a volume of Dick's writings edited by Lawrence Sutin's *The Shifting Realities of Philip K. Dick* (Random House: Vintage, 1995, p. 233).

[3] The book was published on March 31, 2019, exactly twenty years to the day after *The Matrix* was released on March 31, 1999, and the talk was on July 22, 2019. You can find the talk on YouTube on *Talks at Google*.

[4] The famous quote seems to have been ad libbed in Metz and can be seen in the various videos of the talk, but I couldn't find it in the written version. In the past decade, this quote become a touchstone for people who are interested in simulation theory and questioning reality. It has even made its way into popular songs, including Maxthor's "Another World." It was used extensively in the documentary , Glitch in the Matrix in 2021.

[5] Excerpts from my call with Tessa are available in Episode #1 of my podcast, *The Simulated Universe with Riz Virk*, available on most podcasting platforms and at https://simulateduniverse.podbean.com. An edited version of the transcript is available on my medium account: http://www.medium.com/@rizstanford.

[6] In his book, *I Am Alive and You Are Dead: A Journey into the Mind of Philip K. Dick*," Emmanuel Carrère writes extensively about the content and process of writing Dick's first major literary novel, *The Man in the High Castle*.

[7] Actually Tessa said that Philip told that story many times but she couldn't remember if it was a chain light or switch first. She also implied that Philip himself might have told the story with slightly different parameters.

[8] This was also one of the outcomes in Stephen King's time travel novel, *11/22/63*.

Chapter 2: The Mandela Effect—Real or Mass Delusion?

[9] https://mandelaeffect.com/.

[10] Although the term wasn't defined when the first ten seasons of the *X-Files* originally aired, the Mandela effect featured prominently in the *X-Files* revival for an eleventh season in 2018, in the episode, "The Lost Art of

Forehead Sweat," reflecting its popularity as an Internet meme. See also: https://knowyourmeme.com/memes/the-mandela-effect.

[11] An example of a survey article of pictures: https://finance.yahoo.com/photos/40-mandela-effect-examples-blow-192300563/.

[12] One of many examples: https://www.healthline.com/health/mental-health/mandela-effect.

[13] Broome, Fiona, www.mandelaeffect.com/.

[14] https://www.thehistorypress.co.uk/articles/mandela-from-prison-cell-to-president/.

[15] Colts, Eileen, in the book, *Mandela Effect: Friend or Foe?* edited by Cynthia Sue Larson (Florida – 11:11 Publishing House, 2019), p. 17. For more on Eileen's YouTube channel: www.youtube.com/OneEileenColts. Other contributors to the book include Paulo M. Pinto, Shane C. Robinson, and Vanessa VA.

[16] One of many online examples of misremembering the Lindberg baby: https://medium.com/random-awesome/these-5-mandela-effects-will-blow-your-mind-ffdd08eb3731

[17] https://www.reddit.com/r/MandelaEffect/comments/37kieq/billy_graham _death_memories/.

[18] https://mandelaeffect.com/tianamen-square-tank-boy/.

[19] https://www.newstatesman.com/science-tech/internet/2016/12/movie-doesn-t-exist-and-redditors-who-think-it-does.

[20] https://consequenceofsound.net/2017/04/the-mandela-effect-becomes-reality-with-this-scene-from-sinbads-genie-movie-shazaam-watch/.

[21] https://geekinsider.com/the-mandela-effect-the-bible-bad-changes-to-the-good-book/.

[22] Presbyterians are more likely to say "debts", whereas Methodists or Roman Catholics are more likely to say "trespasses." See an explanation here: https://christianobserver.net/the-lords-prayer-trespasses-another-example-of-the-mandela-effect/.

[23] https://medium.com/@nathanielhebert/the-thinker-has-changed-three-times-b2e54db813fa.

[24] https://medium.com/@nathanielhebert/the-thinker-has-changed-three-times-b2e54db813fa.

[25] Alvin Langdon Coburn (1882–1966)."George Bernard Shaw in the Pose of *The Thinker*" (1906).
There are a number of prints of this photograph based on the original 1906 negative. This particular one was from Cornell's Johnson Museum of Art, https://emuseum.cornell.edu/objects/25750/the-thinker-george-bernard-shaw.

26 The original plaster is in the Rodin Museum in France, but there are many copies, cast in bronze or plaster. This one is from Stanford University's Cantor Arts Center, taken by the author.

27 *Journal of Experimental Psychology: Human Perception and Performance,* *https://www.ncbi.nlm.nih.gov/pmc/articles/PMC3543826/.*

28 https://www.mentalfloss.com/article/585887/mandela-effect-examples.

29 https://www.independent.co.uk/news/science/mandela-effect-false-memories-explain-science-time-travel-parallel-universe-matrix-a8206746.html.

30 "Discredit abuse survivors' testimony by inferring that false memories for childhood abuse can be implanted by psychotherapists." https://www.researchgate.net/publication/11693702_Lost_in_the_Mall_Mis representations_and_Misunderstandings.

31 https://www.independent.co.uk/news/science/mandela-effect-false-memories-explain-science-time-travel-parallel-universe-matrix-a8206746.html.

32 https://www.youtube.com/watch?v=2XNJDr2TFDk&t=7s is one of his talks.

33 It would be interesting to determine whether there were any groups of people who remember JFK being assassinated in a city other than Dallas. A quick search on the Internet reveals that there are many who remember the details of the assassination differently (including some who remember seeing a gun in Jackie Kennedy's hand, which obviously didn't happen, at least in our timeline, making it yet another Mandela effect), though not any significant number who remember an alternate timeline where the assassination didn't occur or occurred in a different city.

34 As an aside, the Higgs boson particle was called the God particle, not so much because of its omnipotent nature, but because it was so "goddamned" hard to find.

35 https://www.reuters.com/article/us-science-cern/cern-scientists-eye-parallel-universe-breakthrough-idUSTRE69J35X20101020.

36 https://www.reuters.com/article/us-science-cern/cern-scientists-eye-parallel-universe-breakthrough-idUSTRE69J35X20101020.

37 A spokesperson for CERN told CNBC, when asked about the Trump theory and the use of CERN in various sci-fi novels and films involving alternate timelines: "These imaginative works, inspired by our scientific research, are works of fiction generated to capture the reader or viewer's sense of wonder and should not be confused with the actual scientific research." https://www.cnbc.com/2017/02/22/alternate-realities-and-trump-mandala-effect-and-what-cern-does.html.

38 https://angelsanddemons.web.cern.ch/faq/black-hole.html.

Chapter 3: The Simulation Hypothesis—Do We Live Inside a Video Game?

[39] These definitions are from a good overview of Berkeley's ideas at: https://iep.utm.edu/berkeley/.

[40] If the universe were to exist for infinity, and in one of those clever arguments that involve infinity, then many of these brains would be formed over time; there may be more Boltzmann brains than actual human brains. As a result, it's quite possible that you are a Boltzmann brain, thinking you are reading a book and complete with false memories. This argument has echoes of the simulation argument we will review shortly from Nick Bostrom, in the idea of using statistics to say that it's more likely we are not in a real world. Most physicists dismiss the Boltzmann brain idea out of hand, arguing that it would take longer than the currently known age of the universe for such a brain to come about randomly from quantum fluctuations. Physicist Brian Greene commented that he was confident that he wasn't a Boltzmann brain, but that he couldn't be absolutely sure about this, nor could our theories. https://www.theguardian.com/science/2020/feb/06/brian-greene-theoretical-physicist-interview-until-the-end-of-time

[41] The term "Buddha" literally means "one who has woken up" or "the awakened one": https://www.theguardian.com/science/2020/feb/06/brian-greene-theoretical-physicist-interview-until-the-end-of-time

[42] Bostrom, Nick, *Are You Living in a Computer Simulation?* Published in *Philosophical Quarterly*. (2003) Vol. 53, No. 211, pp. 243–255. [*www.simulation-argument.com*]

[43] Bostrom, *Are You Living in a Computer Simulation?*

[44] To get a good appreciation of each stage, you can read *The Simulation Hypothesis* or some of my online writing, since our progress on this path changes every three months or so, with some new technological development that inches us closer to this important point. Some of my articles, including "How to Build *The Matrix* on Techcrunch and Digital Trends," talk about these stages as well. For the latest, visit my personal website, www.zenentrepreneur.com, or follow me on twitter @rizstanford, or tune into my podcast, *The Simulated Universe*.

[45] Augmented-reality glasses are becoming smaller all the time, and by the time you are reading this book, there will be AR glasses that look like ordinary sunglasses or eyeglasses, like those depicted in the third season of *Westworld*. In that show, a person puts on the glasses and it appears that they are talking to someone in the room with them while the other party might be located somewhere else entirely.

Just as interesting is haptic technology, such as gloves and body suits, which can be used to simulate the feeling of touching an object (such as a cube or a cup) via small electrical signals that the wearables project onto your skin. These are underway already, promising that what happens in virtual environments can be felt, as depicted in *Ready Player One*.

[46] Since *Pong* keeps coming up, after successfully demonstrating the chip could read the brain signals of a pig, Musk's Neuralink showed that a monkey, with the chip implanted, could be taught to play a video game and its brainwaves monitored. Eventually, the joystick was disconnected from the game, and the monkey, oblivious to this fact, would continue to use the joystick. Unknown to the monkey, the chip was reading its brain signals and beaming them to the game engine. Just in 2021, Neuralink demonstrated that its BCI could be used to have the monkey play the game using brain signals alone.

There are two important variations of brain computer interfaces: invasive and noninvasive. Neuralink's chip is an example of an invasive one, whereas companies like Neurable and Muse and openBCI are examples of noninvasive headsets that read the brain signals.

[47] https://www.technobuffalo.com/stephen-hawking-black-holes-harvard.

[48] Hawking was talking about the black hole information paradox, in which information could fall into a black hole and the information that came out of it might be random. Thus, the information falling in was destroyed. Scientists today think they are on the verge of solving the black hole information paradox. (See articles like https://www.quantamagazine.org/the-black-hole-information-paradox-comes-to-an-end-20201029/ - "The Black Hole Information Paradox Nears Its End.") However, I'm using his quote to illustrate the quandary we find ourselves in if we can't be sure of a single fixed past.

[49] For more on the Turing Test, see *The Simulated Hypothesis* or search online. It is a common goal of AI systems, and a number of commentators track whether any AI characters have passed.

[50] Two virtual influencers that have millions of followers as of this writing in 2021 are: Lu do Magalu with more than 14 million followers on Facebook though she is primarily known in Brazil, and Lil Miquela, with over 2 million Tiktok and Instagram followers. Another notable one is Seraphine, based on a character from the popular game, League of Legends and Hatsune Miku, a voice synthesizer who has sung over 100 songs and has millions of followers. These are part of the first wave of virtual influencers, no doubt we will see additional waves that will be more sophisticated virtual characters with AI. One of the early ones is Kuki, originally developed as a chatbot but now with a virtual character attached.

Chapter 4: A Variety of Multiverses

[51] A good summary of both Greene's and Tegmark's classifications is at: https://www.thoughtco.com/types-of-parallel-universes-2698854. Greene

classifies, in *The Hidden Reality*, his parallel universes by name as (1) the quilted multiverse, (2) the inflationary multiverse, (3) the Brane multiverse (related to string theory), (4) the landscape multiverse, (5) the cyclic multiverse, (6) the quantum multiverse, (7) the holographic multiverse, and (8) the simulated multiverse.

Tegmark classifies his multiverse as Level 1 (distant regions of space), Level 2 (inflationary bubbles), Level 3 (quantum mechanics multiverse), and Level 4 (universes based on other mathematical structures). This was done initially in a paper titled "Parallel Universes" for a compilation, *Ultimate Reality: From Quantum to Cosmos*, honoring John Wheeler's 90th birthday. (J.D. Barrow, P. C. W. Davies, and C. L. Harper, eds. Cambridge University Press (2003), available at https://space.mit.edu/home/tegmark/multiverse.pdf. Later, he goes into more details in his book, *Our Mathematical Universe*.

[52] Boiling the types of multiverses down to a finite list is an item of debate among physicists. In this chapter, I have borrowed terms used by Greene, Max Tegmark, and Michio Kaku, in their books: *The Hidden Reality (Greene)*, *The Mathematical Universe (Tegmark)*, and *Hyperspace/Parallel Universes* (Michio Kaku).

[53] This was done using a network of telescopes called EHT, or the Event Horizon Telescope, utilizing a technique called very long baseline interferometry (VLBI).

[54] Penrose diagrams are common online. This one was taken from https://jila.colorado.edu/~ajsh/insidebh/penrose.html.

[55] Greene calls this the quilted multiverse, and Tegmark refers to it as Level I multiverse. They are both the result of eternal inflation.

[56] And you might have thought that most physicists don't believe in magic or God!

[57] "Inflation theory was developed in the late 1970s and early 80s, with notable contributions by several theoretical physicists, including Alexei Starobinsky at Landau Institute for Theoretical Physics, Alan Guth at Cornell University, and Andrei Linde at Lebedev Physical Institute. Alexei Starobinsky, Alan Guth, and Andrei Linde won the 2014 Kavli Prize for pioneering the theory of cosmic inflation" from Wikipedia - https://en.wikipedia.org/wiki/Inflation_(cosmology).

[58] For more details, read Greene, *The Hidden Reality*, pp. 62–67.

[59] Credit: Shutterstock.

[60] Tegmark, *Our Mathematical Universe*, p. 142.

[61] Lemley, Brad. "Why Is There Life?" Discover magazine, reference via Wikipedia.

[62] Tegmark, *Our Mathematical Universe*, p. 142

[63] Schatzman, E. L., & Praderie, F., *The Stars* (Berlin/Heidelberg: Springer, 1993), pp. 125–127, referenced via Wikipedia, entry for "Fine-tuned universe."

[64] https://earthsky.org/space/definition-what-is-dark-energy

[65] Tegmark, *Our Mathematical Universe*, p. 142.

[66] Kaku, Michio, *Hyperspace*, p. 34.

[67] Kaku, Michio, *Hyperspace*, p. 69–77.

Chapter 5: Welcome to the Quantum World

[68] Oxford Languages, 2021

[69]

https://en.wikipedia.org/wiki/Annus_Mirabilis_papers#Photoelectric_effect

[70] https://en.wikipedia.org/wiki/File:Double-slit.svg

[71] https://www.britannica.com/biography/Erwin-Schrodinger

[72] Technically the probability is the square of the amplitude above the seat, but the main point is that the probability wave shows the different amplitudes over different positions in space.

[73] https://commons.wikimedia.org/wiki/File:Particle2D.svg (public domain)

[74] Image by A Friedman, taken from
http://afriedman.org/AndysWebPage/BSJ/CopenhagenManyWorlds.html

[75] Wheeler, who crops up again and again in this book, spent time with the founders of the new quantum theory in Europe before returning the US and becoming, in his later years, the "grand old man" of physics because of his institutional memory of working with Bohr, Heisenberg, Schrodinger, Born, Einstein and many others.)

[76] J. A Wheeler, in the Physicist's Conception of Nature, ed. J. Mehra (Dodrecht, Holland: D. Reidel, 1973) p244

[77] Quoted in Kaku, Parallel Universes, p188

[78] Greene, Brian, The Hidden Reality, p 232

[79] Greene, Brian, The Hidden Reality, p 228

[80] https://www.scientificamerican.com/article/is-time-quantized-in-othe/

Chapter 6: The Quantum Multiverse

[81] A good discussion of the history is given here:
https://www.aps.org/publications/apsnews/200905/physicshistory.cfm#:~:t
ext=May%2031%2C%201957%3A%20DeWitt's%20Letter,in%20its%20own%2
0separate%20universe.

[82] This summary of titles was based on University of California, Irvine's website, http://ucispace.lib.uci.edu/handle/10575/1302. Many places reference the thesis by many titles, and multiple versions (long and short and summary versions) have been published in various physics journals.

[83] https://www.aps.org/publications/apsnews/200905/physicshistory.cfm.

84 https://www.quantamagazine.org/why-the-many-worlds-interpretation-of-quantum-mechanics-has-many-problems-20181018/ (Italics on "possible futures" was added by me.)

85 Deutsch, David (2010). "Apart from Universes." In S. Saunders; J. Barrett; A. Kent; D. Wallace (eds.). *Many Worlds? Everett, Quantum Theory and Reality.* Oxford University Press, as quoted via Wikipedia.

86 https://commons.wikimedia.org/wiki/File:Interference_of_two_waves.svg

87 DeWitt has told the story many times. In this case, I am referencing it from DeWitt's meeting with Tegmark, described on p. 191 of *Our Mathematical Universe.*

Chapter 7: The Nature of the Past, Present, and Future

88 Buonomano, Dean, *Your Brain Is a Time Machine* (New York: W. W. Norton & Company, 2017) p. 4.

89 https://en.wikipedia.org/wiki/Light_cone#/media/File:World_line.svg.

90 These images were taken by Eadweard Muybridge of a horse (owned by California governor Leland Stanford) galloping in 1878, which led to the development of motion pictures.

91 Found in: https://towardsdatascience.com/quantum-gravity-timelessness-and-complex-numbers-855b403e0c2f, based on original found in *The Origin of the Universe*, by John D. Barrow.

92 https://commons.wikimedia.org/wiki/File:Wheeler_telescopes_set-up.svg (Source: Patrick Edwin Moran).

93 https://www.sciencealert.com/wheeler-s-delayed-choice-experiment-record-distance-space.

94 Constructed by the author with clipart from Shutterstock.

95 https://www.wikiwand.com/en/Wheeler%27s_delayed-choice_experiment.

96 Created by author. For more detailed treatment of snapshots, please see *The Fabric of Reality* by Deutsch.

Chapter 8: Multiple Timelines in *SimWorld*

97 Lloyd, Seth, *Programming the Universe*, p. 3.

98 Virk, *The Simulation Hypothesis* (2019), pp. 30–31.

99 https://en.wikipedia.org/wiki/Colossal_Cave_Adventure.

100 "binary" means "binary notation", which is a representation of number using only bits, like this: 1011. It is in contrast to a decimal notation, which uses normal digits (0-9). For example, in binary the number 1 is represented as 0001b, 2 is 0010b, 3 is 0011b, 4 is 0100b and so on.

101 5 rows * 11 aliens in each row = 55 aliens. Since each alien might have 8x8 set of pixels, that would give us 55 * 64 pixels = 3520 pixels. Assuming each

pixel was one bit (i.e., only one color, either it's on or off), you could save the whole gamestate with 3520 bits, or 440 bytes. Rounding up to the nearest power of 2, this means 512 bytes would be more than enough to store the *Space Invaders* gamestate.

[102] https://en.wikipedia.org/wiki/Context_switch.

[103] Source: Shutterstock

Chapter 9: Simulation, Automata, and Chaos

[104] Stephen Wolfram in TED Talks, "Computing the Nature of Everything," 2010, one link: https://www.youtube.com/watch?v=60P7717-XOQ.

[105] Wills Peter R. ,2016,"DNA as information", *Phil. Trans. R. Soc.* A.3742015041720150417, http://doi.org/10.1098/rsta.2015.0417

[106] Levy, *Artificial Life*, p. 31.

[107] Ibid.

[108]This quote is taken from a larger quote by Fredkin in Levy, *Artificial Life*, p. 63.

[109] https://www.latimes.com/archives/la-xpm-2002-jun-09-adna-newsci-story.html.

[110] https://mathworld.wolfram.com/ComputationalIrreducibility.html.

[111] https://en.wikipedia.org/wiki/Reversible_computing.

[112] Amoroso, S. and Patt, Y. N. (1972), "Decision procedures for surjectivity and injectivity of parallel maps for tessellation structures," *Journal of Computer and System Sciences*, 6 (5): 448–464, doi:10.1016/S0022-0000(72)80013-8, MR 0317852.

[113] Source:
Left side: https://commons.wikimedia.org/wiki/File:Rule30-256-rows.png
Right side: https://commons.wikimedia.org/wiki/File:Textile_cone.JPG
8

[114] Source for Sierpinski triangle:
https://commons.wikimedia.org/wiki/File:Sierpinski_triangle_evolution.svg.
The Koch snowflake iterations are in the public domain but available here:
https://commons.wikimedia.org/wiki/File:Koch_Snowflake_0th_iteration.svg
.
https://commons.wikimedia.org/wiki/File:Koch_Snowflake_1st_iteration.svg
.
https://commons.wikimedia.org/wiki/File:Koch_Snowflake_2nd_iteration.sv
g.
https://commons.wikimedia.org/wiki/File:Koch_Snowflake_3rd_iteration.svg
.
https://commons.wikimedia.org/wiki/File:Koch_Snowflake_4th_iteration.svg
.

[115] https://en.wikipedia.org/wiki/File:Mandel_zoom_00_mandelbrot_set.jpg

[116] A more complete and mathematical discussion of randomness in fractals is provided in Manfred Schroeder's *Fractals, Chaos, Power Laws,* and this use case is quoted from p. 30.

[117] This image is was the Hata's tree-like set. Croydon, David. "Random fractal dendrites." PhD thesis, St. Cross College, University of Oxford, Trinity, 2006. Fig 1.5, p64.

[118] *Problème des trois Corps*), based on the competing work of Jean le Rond d'Alembert and Alexis Clairaut in the 1740s.

[119] https://www.scientificamerican.com/article/quantum-mechanics-free-will-and-the-game-of-life/.

[120] Wolfram argues in his paper, "Random Sequences Generated by Cellular Automata," that using a combination of XOR (exclusive Or) and OR gates and mathematical functions, that a random sequence can be generated by CAS.

Chapter 10: Quantum Computing and Quantum Parallelism

[121] Lloyd, Seth, *Programming the Universe*, (New York: Vintage Books/Random House, 2006) p. 7.

[122] There are entire books written as "introduction to programming quantum computers." An example of notation of qubits that starts these books would make even ordinary programmers recoil in horror:

$$(<0| + <1| + <0| + <1|) = \{ \begin{matrix} 1+i & 1-i \\ 1-i & 1+i \end{matrix} \} \frac{\pi}{2} \sqrt[2]{X}$$

[123] For more information on Shannon, I recommend reading *A Mind at Play* by Rob Goodman and Jimmy Soni (New York: Simon and Shuster: 2017)

[124] Feynman, Richard P. "Simulating Physics with Computers," *International Journal of Theoretical Physics*, Vol. 21, Nos. 6/7, 1982. The paper is available many places online; one such location: https://www.cs.princeton.edu/courses/archive/fall05/frs119/papers/feynman82/feynman82.html.

[125] This story has been told in many variations (using rice or wheat). A good overview is given here: http://www.singularitysymposium.com/exponential-growth.html.

[126] Feynman, Richard P. "Quantum Mechanical Computers," *Foundations of Physics*, Vol. 16, No. 6, 1986 , available online; one such location: http://physics.whu.edu.cn/dfiles/wenjian/1_00_QIC_Feynman.pdf.

[127] Brown, Julian, *The Quest for the Quantum Computer*, p. 26.

[128] This definition is an amalgamation of the definition of quantum parallelism from many places. One of these is: https://quantum-algorithms.herokuapp.com/299/paper/node16.html; another is at https://www.sciencedirect.com/topics/engineering/quantum-parallelism. Another good one is in the paper, "Quantum computing methods for

supervised learning," from Viraj Kulkarni1, Milind Kulkarni, Aniruddha Pant, available at https://arxiv.org/pdf/2006.12025.pdf.

[129] Source: https://commons.wikimedia.org/wiki/File:Shor%27s_algorithm.svg

[130] Source: https://commons.wikimedia.org/wiki/File:Quantum_Logic_Gates.png

[131] This gate is named after the French mathematician, Jacques Hadamard (1865–1963), because of a matrix operation called a Hadamard transform, which is often the first step in a quantum operation. We have stayed away from discussions about the mathematics of the various gates, but they are based on representing the input values as matrices and applying other matrices to come up with a mathematical basis for quantum logic gates and quantum computing. The values used in these transforms are based on an abstract sphere, called a Baloch sphere, whose spin or direction indicates the current value of a qubit.

[132] Circuit diagram: https://commons.wikimedia.org/wiki/File:Toffoli_gate.svg

[133] https://www.nytimes.com/2019/05/08/science/quantum-physics-time.html, and the original paper can be found here: https://arxiv.org/pdf/1712.10057.pdf.

[134] Though, of course, you could put a qubit that has been measured back in superposition, using another H gate and measure it again. There is no reliable way to know what the next measurement would be.

Chapter 11: Digital Timelines and Multiverse Graphs

[135] Definitions from Oxford Languages.

[136] In this diagram and in most physics examples, the speed of light is represented by c, as in Einstein's famous equation $E=mc^2$. In this diagram, thus the vertical axis is ct as opposed to t .

[137] https://commons.wikimedia.org/wiki/File:Minkowski_diagram_-_photon.svg.

[138] Technically, those who make microprocessors will tell you that these numbers actually refer to cycles per second, not operations per second. Theoretically, though, each physical cycle of the microprocessor should translate into a kind of operation even if these operations are much lower-level than we are used to thinking of them.

[139] In Wolfram's project, each point in space is defined as a node, and time is defined as the computation that is needed to go from one node to another. Wolfram claims that you can derive the fundamental laws of physics using branchial space and a multiway graph. For example, the number of computations used to get from one point to another is the key to time dilation, which takes on a new meaning in a computational universe. A

person traveling at or near the speed of light only needs to compute one operation while others are computing many more operations.

[140] In our universe, with approximately 10^{80} atoms, each gamestate would be absolutely massive, with each possible string of 10^{80} values representing one node in a giant multiway graph.

Chapter 12: The Core Loop as Search

[141] Hans Moravec, 1998, Simulation Consciousness, Existence, available at CMU
https://frc.ri.cmu.edu/~hpm/project.archive/general.articles/1998/SimConEx.98.html

[142] We can do this by changing the formula to one that is self-adjusting, such as the logistic equation, replacing the formula with $X_{t+1} = rX_t(1-X_t)$. Using this equation, the population growth rate slows down as the population gets large and reaches some maximum value.

[143] author: Nuno Nogueira,
https://commons.wikimedia.org/wiki/File:Minimax.svg

[144] Hoffman, Donald, *The Case Against Reality*, p. 53 (New York: W.W. Norton), 2019.

[145] Wikipedia, https://en.wikipedia.org/wiki/Genetic_algorithm.

Chapter 13: The Upshot—The Universe Evolves Through Multiple Simulations

[146] The Garden of Forking Paths was the first of Borges's works to be translated into English by Anthony Boucher when it appeared in *Ellery Queen's Mystery Magazine* in August 1948. Source for this and other facts: Wikipedia entry for "The Garden of Forking Paths." The story may be read in many places online, including:
http://mycours.es/gamedesign2012/files/2012/08/The-Garden-of-Forking-Paths-Jorge-Luis-Borges-1941.pdf.

[147] Ibid.

[148] Physicists ranging from Michio Kaku to science fiction writers such as Olaf Stapledon and others reference "The Garden of Forking Paths" when trying to explain this idea of multiple timelines.

[149] A word that seems dated even though it's technically still valid, though we are more likely to say, "Chinese studies" today than "sinology."

[150] From Google's DeepMind website.

[151] This is only one of many self-driving simulation systems, described at https://news.mit.edu/2020/system-trains-driverless-cars-simulations-0323.

[152] Thomas Campbell, My Big TOE, (Lightning Strike Books, 2003),201.

[153] I also stumbled at the time across the paper, "On Testing the Simulation Theory"; it is available at cusac.org and was authored by Tom Campbell, Houman Owhadi, Joe Sauvageau, and David Watkinson. This paper described experiments to find evidence that the universe is rendered by a conscious entity, like a player in a video game. The experiments that are described are currently underway at a university in California as I finish writing this book in early 2021, but have reached no definite conclusions yet.

[154] For a full discussion, see Wolf, Fred Alan, *Parallel Universes: The Search for Other Worlds* (New York: Simon and Shuster, 1989) pp. 218–224.

[155] Wolf, Parallel Universe, p 222

Chapter 14: Stepping Back—What Does It All Mean?

[156] Goswami, Amit, *The Self-Aware Universe* (New York: Tarcher/Putnam) p. 140.

[157] https://www.simulation-argument.com/

[158] Sudman, *Application of Impossible Things*, p. 59–60.

[159] Brian Weiss, *Messages from the Masters* (New York: Warner Books, April 2001), p. 45.

[160] Michael Newton, *Journey of Souls*, p. 206 (St. Paul, MN: Llewellyn Publications, 1994), Second Revised Edition, p. 207.

Index

About the Author

Rizwan ("Riz") Virk is a successful entrepreneur, investor/venture capitalist, bestselling author, video game industry pioneer, and indie film producer. Riz was the founder of Play Labs @ MIT (www.playlabs.tv), and is a venture partner at Griffin Gaming Partners. Riz is a graduate of MIT in Computer Science, and Stanford's Graduate School of Business in Management, and is currently at the School for the Future of Innovation in Society at Arizona State University.

Since catching the startup bug at the age of 23, Riz has been a founder, investor and adviser in many startups, including Gameview (DeNA), CambridgeDocs (EMC), Tapjoy, North Bay, Funzio (GREE), Pocket Gems, Disruptor Beam, Discord, Telltale Games, Theta Labs, Tarform, Upland, and 1BillionTech. These startups have created software used by thousands of enterprise and video games with millions of players, such as *Tap Fish*, and games based on *Penny Dreadful, Grimm, Game of Thrones, Star Trek* and *The Walking Dead*.

Riz has produced many indie films, including the online phenomenon *Thrive: What On Earth Will It Take?*, *Sirius, Knights of Badassdom, The Outpost* and adaptations of the works of Philip K. Dick and Ursula K. Le Guin.

Riz is the author of *The Simulation Hypothesis, Startup Myths & Models: What You Won't Learn in Business School, Zen Entrepreneurship* and *Treasure Hunt*. Riz's writing and startups have been featured everywhere from Tech Crunch to The Boston Globe from Vox.com to NBCNews.com, from Coast-to-Coast AM to the History Channel.

He lives in Mountain View, CA, Cambridge, MA and Tempe, AZ. His personal and professional sites are www.zenentrepreneur.com and www.bayviewlabs.com.

DON'T MISS OTHER TITLES BY

RIZWAN VIRK

THE SIMULATION HYPOTHESIS

STARTUP MYTHS & MODELS

ZEN ENTREPRENEURSHIP

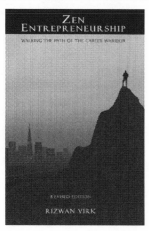

TREASURE HUNT

visit
http://www.zenentrepreneur.com/